at home in the world

at HOME in the WORLD

Human Nature, Ecological Thought, and Education after Darwin

Eilon Schwartz

Published by
STATE UNIVERSITY OF NEW YORK PRESS
Albany

© 2009 State University of New York

All rights reserved

Printed in the United States of America

No part of this book may be used or reproduced in any manner whatsoever without written permission. No part of this book may be stored in a retrieval system or transmitted in any form or by any means including electronic, electrostatic, magnetic tape, mechanical, photocopying, recording, or otherwise without the prior permission in writing of the publisher.

For information, contact
State University of New York Press, Albany, NY
www.sunypress.edu

Production, Laurie Searl
Marketing, Anne M. Valentine

Library of Congress Cataloging-in-Publication Data

Schwartz, Elion, 1958–
  At home in the world : human nature, ecological thought, and education after Darwin / Elion Schwartz.
    p. cm.
  Includes bibliographical references and index.
  ISBN 978-1-4384-2625-9 (hardcover : alk. paper)
 1. Nature and nurture.   2. Evolution (Biology)   3. Social Darwinism.
4. Dewey, John, 1859–1952.   I. Title.
  BF341.S38 2009
  370.1'2—dc22   2008042784

10 9 8 7 6 5 4 3 2 1

This book is dedicated to the memory of

*Tzvi Kress,
my zeidi,*

and to

*Nathan Schwartz,
my father.*

*Their gentleness, generosity, and empathy
for others taught everyone around them about the
depth and breadth of human goodness.*

Contents

Acknowledgments ix

CHAPTER ONE
The Making of Darwinism 1
    Darwin and the Good in Human Nature

CHAPTER TWO
Nature's Lessons: Applying Evolutionary Theory
to Educational Philosophy in the Nineteenth Century 25
    From Philosophy of Nature to Educational Philosophy
    Spencer's Educational Philosophy
    Huxley's Educational Philosophy
    Kropotkin's Educational Philosophy
    Conclusion

CHAPTER THREE
Dewey's Darwinism: Human Nature
and the Interdependence of Life 59
    Change and Growth Are the Essential Features of Darwinism
    Human Beings Can Only Be Understood as Part of the Natural World
    The Natural and the Social: Dewey's Notion of Habit
    Human Beings Are by Nature Social Animals,
        and Can Only Be Understood Through Their Sociability
    Living in the World: Democracy as a Natural Value
    Dewey's Darwinian Educational Philosophy

CHAPTER FOUR
Mary Midgley and the Ecological *Telos*   85
Innate Needs
The Teleological Implications of Having Needs
Feminism and Human Nature: A Case Study in Teleological Thinking
On Building a Whole Life
Moral Objectivity and the Reality of Evil
Breaking Down the Is/Ought Dichotomy
A Transcendent Life

CHAPTER FIVE
A Darwinian Education   129
The Aims and Purposes of Education: A Darwinian Perspective
Emotions and Reason
Particularism and Universalism
From Nature to Culture: A Darwinian Curriculum
Cultivating Wonder: Educational Didactics
Final Thoughts

Notes   167
Index   207

Acknowledgments

The writing of this book spans the early childhood years of Ayelet and Chen, my two oldest children, the birth of the newest addition to the family, Eden, and then some. Ten years of study and its distractions allows you to build a rather impressive list of people and places which will remain forever linked with this journey.

Prof. Avner De-Shalit, my doctoral advisor and more, epitomizes for me the rare but so desperately needed combination of intellectual rigor and human warmth and generosity. It was deceptively easy for me to make the arguments of this book on the fundamental decency of human beings with Avner around. The Melton Centre for Jewish Education in the School of Education at the Hebrew University has been my academic home, and they have been extraordinarily supportive in allowing me to pursue my interests far beyond the usually defined borders of "Jewish education." I had the rare pleasure of spending a sabbatical year at the Global Environment Program at the Watson Institute for International Studies at Brown University, where I worked on the adaptation of my doctoral thesis into a book. Providence is a great little city, Brown a great university, and both a welcome respite from the hustle and bustle of life in Tel Aviv. Laura Sadovnikoff and Prof. Steven Hamburg helped me set up a second home while there, welcoming me into the stimulating Brown and Watson communities. And I am indebted to Dr. Daniel Orenstein for having encouraged and helping to facilitate the year at Brown.

While on sabbatical I took a leave of absence from The Heschel Center for Environmental Learning and Leadership, and my friends and colleagues needed to often work extended hours filling the gap of my absence. A special thank-you goes to Orli Ronen-Rotem, whose generosity of time and spirit allowed me to be away for the year. Dr. Noah Efron of Bar Ilan University critically read the original doctorate on which this book is based, and, in sharing intellectual concerns so very similar to my own, has been a welcome sounding board for working through ideas. And no list of thank-yous is complete without thanking Dr. Jeremy Benstein, my extraordinary colleague and friend for a quarter of a century, whose conversation throughout the years has always served to inspire and challenge my thinking and commitments.

The Dorot Foundation supported my work with a generous fellowship. Dr. Ernie Freirichs did not hesitate for a moment when I approached him as representative of the foundation about possible funding, and his belief in the importance of my academic pursuits was, and continues to be, greatly appreciated. Charlotte Talborth of the Levinson Foundation also helped out with additional funds. The Cummings Foundation helped to support my work during our sabbatical in Providence. Peter Teague, head of Cummings' Environmental Program, enthusiastically extended his help, and it is an honor to have him as a colleague. My family and I will be forever in debt to all of them and their respective foundations for not allowing the project to sink us into a huge financial hole.

My brother shared with enthusiasm his pride in my accomplishments—no easy thing for two boys who grew up on sibling rivalry. I was raised, among other things, on my mother's critical abilities, and the conversations late into the night growing up around the table were undoubtedly one of the original sources of my own critical facilities. Rachel, my wife, while working a full-time job, researching her own thesis, and raising three children, miraculously managed to carve out space for my work, as well.

There are many gifts that my marriage has brought, but in the context of this book, none has been more significant than the three amazing kids that complete our family. Writing about human nature with new life surrounding you is a reliable way of making sure that theory and practice, philosophical and empirical inquiries remain tightly linked. No one was going to tell me that they were blank slates as I saw their personalities emerge at the earliest moments of life, and no one was going to convince me that humans are born selfish as I delighted in the daily affections of fatherhood.

Finally, this doctorate is dedicated to the memory of my maternal grandfather, Zvi Kress—Zeidi, as his four grandchildren affectionately called him from the Yiddish—and to my father, Nathan Schwartz. Although they were not related biologically, they were both living proof of the profound decency of which human beings are capable. Their memories shine for me as a moral beacon as to how to live a life that is infused with a seemingly effortless love, concern and commitment for family and friends, the Jewish people, and, with little exaggeration, for human beings wherever they were met. Beyond the philosophical arguments of this book, I hope that I have conveyed the lessons of this inheritance in the pages before you.

CHAPTER ONE

The Making of Darwinism

The contemporary debate about the nature of human nature, centering around the implications of Darwin's theory of evolution, is the newest chapter in a long history of explorations. Conflicting ideas about human nature have always sat at the core of philosophical debates, often educational ones. Plato and Aristotle, for example, had differing views on human nature, and thus different approaches to educational philosophy.[1] So too did Descartes, Hobbes, Locke, Hume and Rousseau.[2] Whether we are essentially atomistic or social beings; whether we are primarily motivated by selfishness or altruism; whether our rationality is an extension of, or at war with, our emotions—all these are of critical import for defining what education can and should be. However, Darwin's theory of evolution radically changed this historical debate on human nature, offering for the first time an empirical basis for the normative discussion. At its heart, I argue, is a simple but far-reaching insight—we have evolved as a profoundly social species, biologically related to the rest of the natural world, and at home in the only planet in the universe for which we are adapted to live. Such a view of human nature, rooted in our best scientific knowledge, has significant implications for how we think about educational philosophy.

Not all proponents of Darwinism, of course, agree with such an interpretation of the meaning of Darwin's theory of human nature. While debates continue to rage between Darwinists and anti-Darwinists, making headlines most recently with the evolution versus intelligent design debates, internal debates between Darwinists are often no less heated. Many Darwinists have essentially continued what can be called the blank slate worldview, a view which has had tremendous influence on progressive ideas. The most famous of such advocates is Stephen Jay Gould, who argued that, although evolved from the rest of the natural world, our intelligence has largely allowed human beings virtually unlimited plasticity.[3] Differences in society are matters of sociology, not biology. And if human society created the differences between rich and poor, black and white, men and women, human society can also erase those differences. The blank slate position, with roots in John Locke's liberalism, gained much influence in the early twentieth century as a reaction to the use of Darwin's theories to justify supposed innate differences between people. Gould continued the tradition of the blank slate, but explained it in Darwinian terms. He called it 'biological potentiality.' As his ideological compatriots argued, biology could probably only explain the most basic human behaviors of "eating, excreting and sleeping."[4] Education in this view has the power to correct social inequalities, and to help reshape human society based on a chosen set of progressive social values.

Richard Dawkins has served as Gould's foil, and in many ways defines the other pole of the debate. Dawkins' extraordinarily powerful metaphor of "selfish" genes suggests an underlying moral truth about the world, where the innate nature of human beings, like the rest of the natural world, can best be described as selfish.[5] Reaching back at least to Hobbes' description of the ultimately self-serving, aggressive and competitive nature of human life, and resurrecting motifs from popular notions of social Darwinism from the late nineteenth and early twentieth centuries, in which such behavior was celebrated as the motor of progress, Dawkins and his supporters accept social Darwinist descriptions of human nature. However, unlike the social Darwinists, Dawkins holds that our natures, like in Freud's psychology, are not our allies. We need to rebel against our genes if we want to create a humane society, but we always need to know that our efforts will be thwarted by our natures, and therefore we need to have realistic expectations about what is possible.[6] In educational terms, our rationality, while always compromised by our baser motives through rationalizations, is our major, if flawed, tool with which to combat our innately selfish motives. Our ideals are built through rational inquiry, transcendent and in opposition to our natures.

It is interesting and important to notice that, although Gould and Dawkins stand in some sense as polar opposites, their worldviews share one very central and critical characteristic. Both see human ideals as originating through rational thought, which can stand independent of genetic determinism, and which is something uniquely human. For Dawkins, our ideals are a rebellion against an insidious nature; for Gould, there is nothing to rebel against. But for both, the natural world is not prescriptive for human beings.

I argue a different position. Following in the footsteps of Kropotkin, Dewey and the contemporary philosopher Mary Midgley, I hold that, indeed, humans have an innate nature, and that, while not dictating human actions, it shapes them far more widely than Gould would accept. However, unlike Dawkins, I hold that we are not at war with ourselves. We are a coherent species, like all species, and our motives and intellect are integrated, not conflicting, parts of a whole. Nor are we innately selfish beings, in competition and at war with each other. Our innate natures can be trusted as good, although certainly not foolproof, beginner's guides which are shaped through our intellects and cultures, and which lead us outward and help to structure life's meaning.

This book, then, argues for a particular interpretation of Darwin, one which affirms that humans indeed have an innate nature, but that it is largely cooperative rather than competitive, social rather than self-centered, communal rather than atomistic. This is not to say that motives of aggression, selfishness and individualism, for example, are socially constructed and foreign to innate human nature. They too, are part of the human condition. While they can indeed be destructive, as can any motive when it eclipses all others, they are more properly seen as moderated by a complex set of interconnected emotions which emerge in a wide set of human behaviors. Evil is a real possibility, but it is not predetermined by our genes. I believe, as shall become evident, that this is a proper interpretation of Darwin, and that such a position can form the basis for a compelling educational philosophy.

I use the term 'innate' provocatively, but also warily. I am not ignorant of the massive amount of literature which exists, contextualizing scientific theory within social ideas and ideals. As is popularly known, we are all postmodernists now. Still, there is a very large difference between staying aware of the ways in which science and culture interface and deeming all scientific claims to be nothing more than hegemonic ideologies writ in supposedly scientific objectivity. The explanatory power of science is too great to deny its descriptive power of the world. Darwin was certainly a child of his times, and his Victorian ideals, and prejudices, are often painfully present. But his ideas can also transcend his times, even as they are embedded within them.

There are five parts to my argument, divided by chapter. In the first chapter, I look at Darwin's theory, and demonstrate why, although open for interpretation, the seeds for a cooperative view of human beings are planted by Darwin himself. In the second chapter I illustrate first attempts to build an educational philosophy based on such a perception of human nature, concentrating on the earlier attempts by Peter Kropotkin, and then, subsequently, in the third chapter, the far more sophisticated and integrated attempts of Dewey. Kropotkin and Dewey are both examples of what I call first-generation Darwinists. In the fourth chapter, I analyze the work of the contemporary British philosopher Mary Midgley, a second-generation Darwinist, and her expansion of Darwin's intuitions about human nature into a robust view of human nature, and its implications for our connection with the natural and social world within which our lives find and express their meaning. I believe Midgley's philosophy powerfully recasts an Aristotelian worldview out of Darwinian biology, one in which our natures significantly help us to define the good, and move us in its direction. This Aristotelian tendency is present already in Darwin, and is developed by Kropotkin and surprisingly by Dewey. Finally, I look at what a contemporary Darwinian educational philosophy emerging from Kropotkin, Dewey and Midgley's philosophies looks like, a philosophy rooted in our understanding of ourselves as part of, and not apart from, the rest of the natural world.

DARWIN AND THE GOOD IN HUMAN NATURE

> "He who understands baboon would do more toward metaphysics than Locke" (Darwin 16 August 1838. M notebook).

Darwin was in many ways the first Darwinist—that is to say, he understood that his theories on the evolution of the natural world had implications for how we understand human life and its meaning. In *On the Origin of Species,* however, he consciously ignored these implications, save for his closing paragraph's cryptic phrase: "light will be thrown on the origin of man and his history. . . ."[7] Darwin knew that evolutionary theory, when applied to human beings, would explicitly confront contemporary cultural and religious views of human origins and the meaning thereof. By tactically separating the question of evolution from that of human origins and meaning (although it was implicit in his argument), Darwin gave his theory of evolution a greater chance of being accepted. After the general theory gained legitimacy, Darwin assumed, it would be politically easier to address its meaning for human beings.[8]

Darwin's theory of evolution, as articulated in *On the Origin of Species,* was controversial enough. Ernst Mayr shows the ways that evolutionary theory as a scientific claim about the origins and development of the natural world challenged accepted orthodoxies, secular as well as religious.[9]

Darwin's theories, inspired by Malthus's theory of population growth, held that there was an inevitable *struggle for existence* of organisms, owing to the geometric growth of populations versus the arithmetic growth of resources.[10] Since resources cannot keep pace with population growth, eventually there would be a struggle for survival. Organisms of a species *vary* from one another in subtle ways, and those characteristics are *inherited* in the next generation—two facts which breeders had known for millennia. Characteristics which improved the survival skills of the individual would then be successfully passed down to the next generation, whereas those individuals who were less fit would be less successful at surviving, and their traits would be less likely to continue into the next generation. Over extended periods of time the traits that led to increased fitness would spread throughout the population. This change of traits within a population could eventually lead to the evolution of a new species. If populations of a species were isolated from one another, for example, they would change independently of one another, and could eventually develop into distinct species. By showing that species were not stable and timeless entities, but rather constantly in flux, Darwin was changing the concept of species from a static one to a dynamic one, in which the boundaries between species are matters of degree, and not kind. Darwin called this process leading to species change and speciation *natural selection*. While breeders had purposefully selected traits among a species, here nature was doing the selecting. The power of the theory is in demonstrating a material mechanism that could explain how species change and can, over geologic time, evolve into new species.[11]

Darwin clearly stated that he used the concept of the struggle for existence "in a large and metaphorical sense."[12] It is often a struggle between organisms of a species, where one organism is more successful than others owing to certain physiological or character traits. This is the accepted meaning of struggle for existence, which Herbert Spencer popularized in his phrase "survival of the fittest," but Darwin pointed out that it is not the only way that the struggle takes place.[13] The struggle for existence can also be a struggle for survival against the natural elements. Particularly in the extreme environments of deserts, mountains and tundra, for example, the struggle is in finding ways of exploiting nature successfully in order to survive.[14] Strategies of cooperation among individuals are particularly successful in such climates. Peter Kropotkin, the Russian prince, anarchist and first-generation Darwinist (that is, those applying Darwin's scientific ideas into its social implications in the years immediately after the publication of Darwin's theories), gave special attention to this form of struggle for existence, due to his natural history education in the extreme climate of Siberia, and this significantly influenced his views on the meaning of nature and human nature, as we shall see. Darwin however, while maintaining that there are more possibilities for evolutionary struggle than that between individuals, nevertheless held that the struggle between individuals of the same species is the most significant and widespread variety of struggle that one actually finds in nature.[15]

The distinction is an important one. Although Darwin saw actual struggle between individuals as the primary strategy for surviving in the natural world, Darwin acknowledges that it is but one strategy, and that cooperation, for example, is an alternative strategy for survival. In other words, although 'struggle' as metaphor describes what individuals and species do, there are multiple strategies that can work. Struggle between individuals is not the only alternative open to species. For Darwin, as we shall see, cooperation is in fact the dominant strategy of the human species.

Darwin's use of the term 'struggle' ultimately did not apply to simple physical survival. From an evolutionary perspective survival is not an ends in itself, but rather a necessary although not sufficient condition for having descendents. Longevity is only important insofar as allowing a long enough lifetime to have offspring, in order to pass one's characteristics down and have them spread through subsequent generations. An organism whose life span might be several hours, but who produces thousands of offspring, many of whom survive to themselves reproduce, is more successful from an evolutionary perspective than an organism of the same species which lives longer but produces fewer descendents who can survive and reproduce. Natural selection is not purposeful. Its sole criteria are the effects of physiological or character traits on the success of the organism to pass on its features to the next generations. Traits which contribute to such success relative to other traits gain precedence, since they give the organism a competitive advantage and thus a greater chance of survival and reproduction.

Because successful strategies of survival are ultimately linked to questions of reproduction, Darwin spoke of sexual selection as a secondary filter for evaluating traits.[16] Both females and males have an evolutionary interest in finding a mate with desirable traits to pass on to their prodigy, including traits which will secure them an evolutionarily desirable mate. This can lead, for example, to the selection for physical strength, a potentially desirable characteristic for winning a sexual partner. In bird species, Darwin showed that males often advertise their physical strength through peaceful means—such as being the most colorful, or having the most attractive birdsong. Darwin recognized that, because of the centrality of reproduction in natural selection, the competition for sexual partners was critical and would lead to different sets of traits being advantageous to males and females, with clear implications for innate differences between men and women.[17] Darwin's attempts to describe these differences were not his best moments, as we shall see.

Darwin did not see natural selection as the sole means of modification of the species. Although many later Darwinists, principally Dawkins and his advocates, see natural selection as the almost exclusive way in which species evolve, Darwin allowed room for other mechanisms, as well. He recognized that there might be other factors involved in evolutionary change, and expressed frustration that he was interpreted as arguing that only natural selection could explain the development within species, and of new species.[18] He did, however, see it as the most critical and dominant factor.[19] This point is central in contemporary Darwinian debates as well. Gould, in opposing Dawkins, attempted to weaken the explanatory power of natural selection. The implication of acknowledging additional mechanisms is that not all characteristics of a species can therefore be explained through their contribution to a species' fitness. Since for Dawkins, natural selection is the predominant explanation for species' characteristics, all characteristics are shaped by their contribution to survival, which Dawkins ultimately describes as a competitive, selfish process. Darwin's position clearly does not support Dawkins' view, here and elsewhere.

One of the most misunderstood components of Darwin's theories, one with perhaps the most radical of implications, was the blindness of the process of natural selection. Evolution was not, in Darwin's view, a slow, steady climb, as argued by a disciple of Herbert Spencer, from gas to genius.[20] The traditional view of the Middle Ages had been of the great chain of being heading linearly downward from God and the angels to the animals and the plants, and from the animate to the inanimate, with human beings located "a little lower than the angels."[21] Advocates of evolution, both before and after Darwin, reversed the direction. It was a steady ladder of progress, with human beings representing the pinnacle of evolution.[22] Such a view does not seem to conform with Darwin's view of the process of evolution, with random variations of character traits occurring between generations, selected by nature according to their relative contribution to survival and reproduction.

On the Origin of Species was published in 1859. It immediately ignited a debate about its implications, but was also debated within the scientific community. Within a decade Darwin felt that it had sufficiently established itself as a credible theory and had been widely adopted.[23] At that point, Darwin was ready to deal with the implications of his theory for human beings. The publication of *The Descent of Man* in 1871 was Darwin's foray into the danger zone of human origins, nature and meaning. After the acceptance of his theory of natural selection spread, Darwin felt confident addressing the issue: "in consequence of views now adopted by most naturalists, and which will ultimately, as in every other case, be followed by other men, I have been led to put together my notes, so as to see how far the general conclusions arrived at in my former works were applicable to man."[24]

Darwin focused his work on showing that there is no boundary of significance between human beings and the rest of nature: human beings are not different in kind from the rest of the natural world. All human characteristics can be found in other species. Curiosity, imitation, attention, memory, imagination, a sense of wonder, reason, progress, toolmaking, language and self-consciousness can all be found in the natural world, particularly among other social animals.[25] Darwin argued that humans, like other species, are different from the rest of nature in degree, but not in kind. All species are different from one another, and all have unique properties, but that doesn't mean that they are not part of the same evolutionary story, sharing many common traits on which their uniqueness is built.

No characteristic of human beings seemed to Darwin more suggestive of the illusory gap between humans and the rest of the natural world than morality:

> I fully subscribe to the judgment of those writers who maintain that of all the differences between man and the lower animals, the moral sense or conscience is by far the most important. . . . It is summed up in that short but imperious word ought, so full of high significance. It is the most noble of all the attributes of man, leading him without a moment's hesitation to risk his life for that of a fellow-creature; or after due deliberation, impelled simply by the deep feeling of right or duty, to sacrifice it in some great cause.[26]

Morality was the "ought" of society. Although morality has been largely described as emerging from our rationality, and rationality is often seen as that which most distinguishes us from the rest of the natural world, Darwin believed that the emergence of most characteristics of a species are explained through the mechanism of natural selection. Morality was a central characteristic of human societies. Being able to explain the emergence of morality in evolutionary terms would show that even the loftiest of human characteristics is rooted in the story of evolution. Darwin was arguing that what societies have come to view as moral behavior has been shaped by natural selection. The "is" of natural selection and its products could explain the origin, and perhaps content, of the "ought" of human morality.

Darwin believed that morality had its basis in social instinct, and that the social instincts were an evolutionary development which instinctively motivated individuals of the species to live in a group, which would give them evolutionary advantages for survival.[27] Like hunger, which developed as an instinct to induce eating (those animals which felt the instinct of hunger were more likely to eat, and therefore had a competitive advantage over individuals who did not feel hungry and would thus presumably eat less), so too the social instinct developed to induce group living. Group living was a successful evolutionary strategy, and therefore social instincts, which encouraged and maintained group living, became favored through natural selection. These social instincts supported a certain personality type, essentially common to all social animals: "they would have felt uneasy when separated from their comrades, for whom they would have felt some degree of love; they would have warned each other of danger, and have given mutual aid in attack or defense. All this implies some degree of sympathy, fidelity and courage."[28] The evolutionary advantage of cooperation with its supporting characteristics emerged out of evolution. From natural selection, which speaks of competition as a mechanism, behaviors of cooperation can develop in species. The human species' strategy of survival was one of cooperation based on sympathy and mutual aid.

Still, it is not clear how Darwin's principles of natural selection could explain the evolution of the moral instincts of sympathy, fidelity and courage from the rudimentary social instincts. If, for instance, in human evolution, individuals living in a group would display acts of courage in battles with neighboring tribes, they would be the most likely to be killed, and the least likely to survive. Properties which might benefit group welfare, therefore, would seem to be selected against, whenever the interest of the individual conflicts with the interest of the group. As most contemporary Darwinists would argue, selection takes place at the level of the individual, making it extremely improbable for behaviors to develop which are beneficial to the group, but detrimental to the individual. Contemporary Darwinists have strengthened the rule, by showing how altruism could develop, as benefiting the group can be a successful survival strategy for the individual.[29] Group selection theory, however, argues that it is possible that at times attributes will be selected which damage the individual's fitness, but increase the fitness of the group. Although largely discredited today, Darwin made a case that this indeed is what takes place:

> It must not be forgotten that although a high standard of morality gives but a slight or no advantage to each individual man and his children over the other men of the same tribe, yet that an increase in the number of well-endowed men and advancement in the standard of morality will certainly give an immense advantage to one tribe over another. There can be no doubt that a tribe including many members who, from possessing in a high degree the spirit of patriotism, fidelity, obedience, courage, and sympathy, were always ready to give aid to each other and to sacrifice themselves for the common good, would be victorious over most other tribes; and this would be natural selection. At all times throughout the world tribes have supplanted other tribes; and as morality is one element in their success, the standard of morality and the number of well-endowed men will thus everywhere tend to rise and increase.[30]

Darwin never articulated how large a role such group selection, as opposed to individual selection, could play. Darwin's choice of explaining natural morality through group selection is from today's evolutionary perspective problematic; individual selection has primarily been the source for evolutionary explanations of morality in second-generation Darwinism However, while group selection allows for behaviors which can benefit the group while being detrimental to the individual, contemporary theories of individual selection strongly support the idea that behaviors which benefit the group can often benefit the individual, as well, and thus there is often no contradiction between the two.

Darwin also understood that altruism could be beneficial to the individual, and could therefore be explained at the level of the individual. Darwin believed that reason would allow individuals to understand that if one aided a fellow creature, s/he would be more likely to aid in return, what in contemporary Darwinism became known as reciprocal altruism.[31] Individual creatures could then learn that cooperation was to their benefit. Although their social instincts enabled the rudimentary motivation to aid another, probably through sympathy, reason reinforced the instinct and elaborated upon it.

The combination of social instinct and reason was not limited to human beings. Darwin's *Descent of Man* was not only about the descent of humans into the natural world and humans being continuous with the rest of nature, but was also about the ascent of the natural world into the privileged place of humans, by showing nature to be continuous with humans.[32] Darwin's *The Expression of the Emotions in Man and Animals*, published a year after *Descent of Man*, demonstrated how much human emotions and animal emotions were alike, and how much human behavior, like that of other species, was universal and innate.[33] While Darwin demonstrated that humans were not only rational, but also instinctual beings, he conversely showed that social animals could do rudimentary reasoning and learning. In fact, Darwin saw the combination of social instincts with reason as a necessary result of evolution that would result in the development of morality in any creature:

> The following proposition seems to me in a high degree probable—namely, that any animal whatever, endowed with well-marked social instincts, would inevitably acquire a moral sense of conscience, as soon as its intellectual powers had become as well developed, or nearly as well developed, as in man.[34]

Since human nature is particular to humans, and human morality emerges out of a particular human nature, it stands to reason, according to Darwin, that different social animal natures would evolve different moralities. Other social animals with advanced reasoning abilities would also develop morality, but the moral codes would be different as a result of their differing natures. If, for example, bees had the same reasoning facilities as humans, bee moral codes might claim that unmarried females sacrificing for the good of the community is their moral responsibility, which would indeed be the natural and moral thing to do.[35] For Darwin, moral behavior was embodied in the nature of the species, and not imposed on the natural world as something foreign to it.

In the debate about the origins and nature of morality, Darwin clearly sided with Hume.[36] Morality was not a derivative of the social contract, or of intellectual reasoning about the greatest good for the greatest number, or about basic human rights derived from a categorical imperative, but was first and foremost a primary instinct rooted in the nature of humans as social, reasoning animals: "the social instincts, the prime principle of man's moral constitution, with the aid of active intellectual powers and the effects of habit, *naturally* lead to the golden rule (emphasis mine)."[37] The golden rule was not a function of rationalizing self-interest, but of instincts, evolved through evolution, being expanded and acted upon.[38] Humans are not sacrificing their natures when they act morally; they are responding to them.

This seems to suggest that morality is not a matter of human choice, but is based on motives programmed into our instincts. Why then do we not live in a perfect world, where each individual is programmed to behave morally by his/her innate social instincts? For Darwin, evolution created conflicting instincts in species. Species are not perfectly integrated beings. Evolution shapes coherency enough to allow the species to survive and reproduce. As an example of conflict between instincts, Darwin discussed female birds at migration, who abandon their hatchlings in their nests in order to migrate, thereby abandoning their maternal instinct for the temporarily stronger migratory one. If birds had developed advanced abilities to reason, Darwin mused that they would no doubt feel enormous guilt when the more acute instinct subsided, and the eclipsed maternal instinct was felt again.[39] Darwin hypothesized that guilt and conscience were a result of reflecting on conflicting instincts, and trying to negotiate between them. He argued that the social instincts were always present, but often weaker than other instincts which could be felt more acutely. When acute instincts, such as self-preservation, hunger, lust and vengeance are acted upon but eventually subside, people are left with their social instincts of sympathy and compassion, and regret having made the wrong choice. For Darwin, this is the evolutionary birth of conscience.[40] Our moral instincts, while often eclipsed by stronger instincts, are more general in our character, and remain after the immediacy of these acute instincts fades away. This is a theme which Midgley develops extensively.

Conscience ultimately acts to temper the desire to act on stronger instincts rather than the weaker, although more persistent, moral ones. Conscience leads humans to develop personal and cultural habits which reinforce the moral instincts and allow them to dictate behavior at moments of conflict. Examples might be cultural and religious moral codes. These codes allow human society to develop to a state where there is no significant conflict between one's habitual instinctual behavior and what is deemed morally right: "Man thus prompted, will through long habit acquire such perfect self-command, that his desires and passions will at last instantly yield to his social sympathies, and there will no longer be a struggle between them."[41] For Darwin, the consummate moral individual is not one who conquers his/her base inclinations at the moment of choice, but rather one who does not experience them.

This shares much with Aristotle's position. Aristotle argued that the virtuous person eventually experiences pleasure at doing the right actions, and pain when doing the wrong ones.[42] For Aristotle, our innate natures suggest a golden mean, which is cultivated through education of the virtues. The virtuous man has trained himself so that there is no felt tension between what one wants to do and what one ought to do. In general, it shouldn't be surprising that Darwin's perspectives often reflect an Aristotelian sensibility.[43] Aristotle, after all, was the philosopher who most systematically integrated a biological understanding into his philosophy.

So, for Darwin, refraining from acting on one's immoral impulses is not enough. Self-restraint is simply selfishness that has been repressed because of fear of societal punishment.[44] Morality is judged by the heart, and not only by deed. It is what motivates the action, and not only the action itself, which determines whether it should be understood as moral. The act is a natural extension of the motive, and cannot be separated from it.[45] An individual who had no moral instinct at all would be "an unnatural monster."[46] Indeed, Darwin claimed that some of the worst criminals have apparently been found to be without any conscience whatsoever, that is, without a moral instinct which stands in tension with baser, stronger ones.[47] There is something unnatural in the perpetrator, something missing from his/her nature, making him/her, on some level, inhuman. Their only instinctual motives are those that derive from serving their own needs, narrowly defined by the acute instincts. Those motivations which reach beyond the individual—of cooperation, sympathy, fidelity, and courage—and which are unique to the social animals, were absent.[48] There is a natural history to both morality and to evil.

For Darwin, morality was the antithesis of selfishness, not a self-serving by-product. He criticized utilitarian philosophers, whom he associated with a "morality born out of selfishness" position, on two counts. First, since Darwin believed that the motives of social animals are what gave birth to morality, and selfishness is a behavior with roots in non-social, and even anti-social behavior, then selfishness cannot be the source of morality, "unless indeed the satisfaction which every animal feels when it follows its proper instincts, and the dissatisfaction felt when prevented, be called selfish."[49] Darwin attacked Adam Smith, for example, for believing that so-called altruistic acts are in fact selfish ones. Smith argued that seeing others suffer reminds us of our own past pains, and therefore relieving their suffering is primarily to relieve our own vicarious suffering.[50]

Secondly, both the selfishness position and the "greatest happiness principle" (which Darwin saw as a later development of utilitarian philosophy) share a view of morality rooted in consciousness and rationality, rather than instinct. The evaluation of whether any particular act increases pleasure or pain demands a level of conscious evaluation which Darwin held is not consistent with the instinctual nature of many moral acts:

> But man seems often to act impulsively, that is from instinct or long habit, without any consciousness of pleasure, in the same manner as does probably a bee or ant, when it blindly follows its instincts. Under circumstances of extreme peril, as during a fire, when a man endeavours to save a fellow-creature without a moment's hesitation, he can hardly feel pleasure; and still less has he time to reflect on the dissatisfaction which he might subsequently experience if he did not make the attempt. Should he afterwards reflect over his own conduct, he would feel that there lies within him an impulsive power widely different from a search after pleasure or happiness; and this seems to be the deeply planted social instinct.[51]

Darwin clearly and passionately rejected the widespread liberal notion of rational self-interest motivating human beings. This is particularly interesting, given the fact that Darwinism is so often seen as offering a biological justification for selfish behavior. Social Darwinism is historically seen as a philosophy which justified selfish behavior as the path to progress. Richard Dawkins has been particularly successful in describing selfishness as the underlying explanation for behaviors in the natural world. The idea that humans, and indeed all species, are at their core selfish, is often attributed to Darwin; it is an idea which is widespread throughout culture and it is fundamentally foreign to Darwin's view.

So, if morality is largely an intuition, what was the role of reason and rationality in Darwin's view? For Darwin, rationality emerges from the evolutionary story, and is linked to sociability and morality. Social species emerged out of the evolutionary process. Learning reinforced social and cooperative intuitions, but did not create them. Moreover, reason allowed the social species to develop a conscience, which also had an evolutionary advantage in allowing the individual to resist selfish behavior, and to pursue the weaker, but more persistent, social behaviors. Here there is choice. Rationality plays a role in negotiating between conflicting urges. The social instincts push in a moral direction, but the individual still needs to choose between opposing options. Rationality is critical to that process.

Ultimately, however, Darwin blurred the distinction between instinctual and chosen behavior, believing that, over time, learned behavior can be inherited:

> . . . as the reasoning powers and foresight of the members became improved, each man would soon learn from experience that if he aided his fellow-men, he would commonly receive aid in return. From this low motive he might acquire the habit of aiding his fellows; and the habit of performing benevolent actions certainly strengthens the feeling of sympathy, which gives the first impulse to benevolent actions. Habits moreover, followed during many generations probably tend to be inherited.[52]

Notice the synergy between the different mechanisms. The feeling of sympathy is already present as a social species, but is reinforced initially by the "low motive" of helping others in order to have the favor returned. The combination of these two, one innate and one chosen, ultimately coevolved so that the selfish motive of helping others in order to help oneself in fact strengthened the sympathetic instinct. Over time the habit of acting benevolently, motivated by both selfish and selfless interests, will itself become inherited, and therefore instinctual.

Darwin believed that over generations learned habits could evolve into a moral instinct, a view that suggests Lamarck's theory of inherited characteristics, which Darwin's theory of natural selection is historically seen to have replaced. Lamarck believed that physical or character traits developed during a lifetime, in interaction with the environment, could be inherited into the next generation. The paradigmatic example of Lamarck's theory is the enormous length of the giraffe's neck, which Lamarck held had stretched a tiny bit in order to reach food in the higher branches of the tree, and the now longer neck was inherited by the giraffe's prodigy.[53] Darwin, of course, was unfamiliar with genetics, and remained unclear as to the biological mechanism of inheritance selection. Mendel's experiments with pea plants, although published before the writing of *The Descent of Man,* nevertheless remained unknown in the scientific community, and only became part of the larger scientific debate at the beginning of the twentieth century.[54] Darwin had his own estimation as to how the mechanism of biological inheritance worked, and at times Lamarckian explanations supplemented them.[55] According to Darwin, learned moral behavior could, over time, become an instinctual habit.[56] He never explained how this would take place, although John Dewey, for example, elaborated on such a process, and built his Darwinism around the concept of habit. Darwin remained pluralistic, assigning the origin of morality in humans to a series of causes: "actions of a certain class are called moral, whether performed deliberately after a struggle with opposing motives, or from the effects of slowly-gained habit, or impulsively through instinct."⁵⁷ He never prioritized between these three categories, and the debate continues to this day. For example, a neo-Lamarckian position which suggests that instincts can be learned through the development of habit suggests a complex nature–nurture dynamic, where morality, perhaps rooted in innate nature, nevertheless is developed and shaped by culture, which through formed habits can then ultimately influence innate human nature.

In Darwin's broadest presentation of the evolution of morals, he states:

> Finally, the social instincts, which no doubt were acquired by man, as by the lower animals, for the good of the community, will from the first have given to him some wish to aid his fellows, and some feeling of sympathy. Such impulses will have served him at a very early period as a rude rule of right and wrong. But as man gradually advanced in intellectual power and was enabled to trace the more remote consequences of his actions; as he acquired sufficient knowledge to reject baneful customs and superstitions; as he regarded more and more not only the welfare but the happiness of his fellow-men; as from habit, following on beneficial experience, instruction, and example, his sympathies became more tender and widely diffused, so as to extend to the men of all races, to the imbecile, the maimed, and other useless members of society, and finally to the lower animals, so would the standard of his morality rise higher and higher.⁵⁸

Darwin assumed that morality would progress, and, in keeping with Victorian sensibilities, believed that Western standards represented the pinnacle of human moral culture.⁵⁹ His Victorian sensibilities are also painfully evident when he applies evolutionary theory to a discussion of differences between men and women, a necessary distinction for Darwin, who noticed the prevalent differences between males and females throughout nature, as sexual selection predicts. Since Darwin held that the human being is, first and foremost, a product of the same evolutionary processes, it stands to reason that similar differences would be found between males and females of the human species:

> With respect to differences of this nature between man and woman, it is probable that sexual selection has played a highly important part. I am aware that some writers doubt whether there is any such inherent difference; but this is at least probable from the analogy of the lower animals which present other secondary sexual characters. . . . Woman seems to

differ from man in mental disposition, chiefly in her greater tenderness and less selfishness; . . . Woman, owing to her maternal instincts, displays these qualities towards her infants in an eminent degree; therefore it is likely that she would often extend them towards her fellow-creatures. Man is the rival of other men; he delights in competition, and this leads to ambition which passes too easily into selfishness. These latter qualities seem to be his natural and unfortunate birthright. It is generally admitted that with woman the powers of intuition, of rapid perception, and perhaps of imitation, are more strongly marked than in man; but some, at least, of these faculties are characteristic of the lower races, and therefore of a past and lower state of civilisation. . . . [60]

While Darwin's descriptions of differences between men and women are all too reminiscent of Victorian ideas which continue to influence our cultural ideas today, many Darwinists, particularly contemporary Darwinists, have also ventured into this extremely contentious discussion, and made similar biological claims. I relate to this debate and try to present a more nuanced position in Chapter Four, through my analysis of the work of Mary Midgley, who, as a feminist, nonetheless argues for acknowledging and embracing innate differences between the sexes, which go beyond biological reproduction systems. One can see here in Darwin's presentation the classic dichotomy with, on the one hand, associating nature with emotions, indigeneous peoples and with women, and, on the other hand, associating culture with rationality, European culture and with men.[61] While rejecting such an equation, I nonetheless advocate for acknowledging and even embracing character differences between men and women, rooted in biology.

Although Darwin clearly celebrated Victorian values as an extension of evolutionary principles, he also believed that in the same way that morality could evolve based on laws of natural selection, its continued development could eventually conflict with group survival. Here we see Darwin's darkest side. For Darwin, a morality based on sympathy for the other would lead to concern for "the imbecile, the maimed, and other useless members of society." Too broad a concern for their welfare would ultimately weaken the group, rather than strengthen it. A gap therefore develops between cultural evolution, which is evaluated according to its moral character, and natural evolution, which is evaluated according to its ability to survive and reproduce, using whatever means necessary. Human ethics and evolutionary ethics part ways. While morals and ethics have their origins in the evolutionary story, their continued pursuit threatens group survival, as others with less concern for the weak gain an evolutionary advantage. Advanced culture, Darwin argued, is in danger of losing its evolutionary fitness:

> ... if the various checks do not prevent the reckless, the vicious and otherwise inferior members of society from increasing at a quicker rate than the better class of men, the nation will retrograde, as has occurred too often in the history of the world. We must remember that progress is no invariable rule.[62]

He suggested that for cultured society to remain evolutionarily fit, and one supposes to be able to remain strong against threats from "the uncivilized," steps need to be taken so that the gap between evolutionary ethics and human ethics does not lead to the fall of civilization and its advanced morality.[63] While recognizing the tension with moral progress, Darwin hinted that the solution to this loss of evolutionary fitness is to restrain moral progress, and to use artificial selection to keep civilization fit, thus foreshadowing and creating the scientific justification for the birth of eugenics in the late nineteenth and early twentieth century:

> With savages, the weak in mind and body are soon eliminated; and those that survive commonly exhibit a vigorous state of health. We civilized men, on the other hand, do our utmost to check the process of elimination; we build asylums for the imbecile, the maimed, and the sick; we institute poor laws; and our medical men exert their utmost skill to save the life of every one to the last moment. ... No one who has attended to the breeding of domestic animals will doubt that this must be highly injurious to the race of man. It is surprising how soon a want of care, or care wrongly directed, leads to the degeneration of a domestic race; but excepting in the case of man himself, hardly anyone is so ignorant as to allow his worst animals to breed.[64]

A morality of supporting the weak while selecting individuals strong in social instincts which are critical for group cooperation and strength, simultaneously keeps weaker members of the group alive, and therefore weakens the groups' physical strength and ultimate survival. Darwin is contradictory about whether one should pursue a natural ethic or a human one. On one hand, he implied that steps should be taken against the biological proliferation of "inferior members of the society"; and yet, on the other hand, he argued that a morality of sympathy for the weak should outweigh the "contingent benefit" of improved biological fitness for the group.[65]

It is helpful to look at Darwin, and in fact all Darwinists, by looking at how they respond to three key questions that can be thought of as three concentric circles. The first, innermost circle is the implication of Darwin's theory for how we understand the natural world apart from human beings. For example, is Tennyson's famous passage—"nature red in tooth and claw" an accurate description of the way the world works?[66] Or is cooperation a more fitting metaphor than competition for the workings of the world? This kind of question sets the stage for the second circle of issues, which is whether this description of the natural world also describes human beings. Since Darwin's theory is popularly understood as tearing down the barricade between human beings and the rest of nature, the argument as to the character of nature is often also an argument about the character of human beings. For example, if there are clear differences in character between the sexes in species, does that mean that there are such differences between the sexes in human beings as well? It matters to education if boys and girls are different in personality because of biology. It matters to education if people are innately aggressive, or if they are socialized to be aggressive. The third circle from which the implications of Darwinism can be viewed is the normative one. Should the description of what is our biologically given human nature have implications for what human beings should be? Descriptive notions of human nature have historically been seen to have normative implications for society, and often, therefore, for educational goals. Rousseau's educational philosophy, for example, was rooted in his view of natural man. His view of human nature not only described the framework in which education works, but it also prescribed what education should strive to be, based on his understanding of human nature.[67] Much of the argument in the second circle, in fact, is predicated on its implications in the third circle. Arguing that boys and girls have no significant innate differences in character and abilities prevents having to confront whether those differences should be turned into a vision for how society should be. Each previous circle's argument is simultaneously about the argument to which it potentially moves. For example, the question about aggression between bands of chimpanzees is simultaneously and inevitably a discussion about whether it is possible for there to be non-aggressive human societies.[68] As we will see, others maintain that the circles are self-contained, that the boundaries are impermeable, and that the answers to an inner circle's questions do not have implications for the outer circle(s). For example, for those that hold that human beings are different in kind from the rest of the natural world, the membrane of the nature–human divide which separates the first and second circle is impermeable. Piaget, for example, believed that humans are unique due to their rationality, and so there was nothing we could learn about human behavior and motives by looking at our animal cousins. For Piaget, the first circle has no implications for the second circle.[69] For those that hold that the "is" of the world has no implications for the "ought" of the world, the membrane of the is/ought divide is impermeable, so that the second circle has no implications for the third. For example, a common strategy in the struggle between science and religion is to claim that science is about facts, religion is about values, and there is no connection between facts and values, between "is" and "ought," and therefore no conflict between science and religion. For such "separatists," the second circle can have no implications for the third.[70]

In analyzing Darwin's Darwinism according to these three concentric circles, several points should be emphasized. In the first circle, in examining how Darwin describes the natural world, it is clear that it is not, as Tennyson proclaimed, only "red in tooth and claw." While the mechanism of competition between individuals within the species was central to explaining how new traits and new species could evolve, at

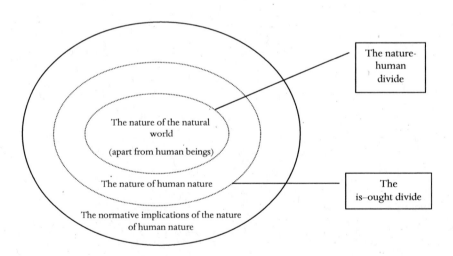

From Darwin to Darwinism—Analytic Circles and Boundaries

[Drawing: The inner circle relates to the debate on the nature of the natural world. If humans are similar to the rest of nature, and therefore it is possible to cross the nature–human divide, then the nature of the natural world will have significant implications for the nature of human nature. If who we are (the "is") has implications for who we should be (the "ought"), and therefore the is–ought divide can be crossed, then the nature of human nature will have significant normative implications for human life.]

times such metaphoric competition could lead to the development of cooperative characteristics in individuals. The social species developed such characteristics, and their nature, therefore, is far from the bloody stereotype of the natural world. Love, sympathy, loyalty and care are all central emotions in the social species of the natural world long before the arrival of human beings.

In the move from the first to the second circle, Darwin's Darwinism is most forceful. There is no significant separation between humans and the natural world. There is a difference of degree, but not of kind. Traits that are considered central to the rest of the animal world are found to be central to humans as well, and traits that are central to humans are found in the rest of the animal world. Emotions are central to human life, as they are to other animals. Rationality and consciousness are not a separate human domain. There is continuity with the rest of the natural world. To prove his point Darwin gave a natural history of morality. He attempted to show that morality, which is considered the most human of characteristics, is a product of the evolutionary process. It can be explained as the result of the emergence of a social species which, through natural selection, has developed an ethic of cooperation that allows the individual to be sympathetic and concerned about the welfare of others. However, he also moved seamlessly into the second circle, not accepting that, even if one accepts the continuity of the human species with the rest of nature, there are unique features of the human species—reason, language, culture—which make it difficult to apply truisms of the natural world so unproblematically to humans. When Darwin spoke about the innate differences between men and women, for example, his methodology suggests that he really did think we are just another species, and that culture does not substantially change our terms of engagement.

The third circle is problematic. Darwin's movement from the "is" of nature to the "ought" of society applies the laws of evolution to the laws of society. It is a methodology that others have followed, and the results have often been disastrous. Darwin himself never described the natural world as one solely dominated by competition and selfishness, but when applying evolutionary theory to human society, he nonetheless often went down the slippery slope of viewing evolutionary success in a competitive world as the standard by which to judge human society. While morality originates within the natural world, its continued extension to all members of society risks undermining the evolutionary health of contemporary culture. Too many invalids protected by well-meaning culture will weaken the 'fitness' of the culture and lead to its extinction.

Darwin, while developing the ideas which allow us to think of ourselves as natural beings, also developed the ideas necessary to justify social engineering based on evolutionary principles. This is an idea which Darwin's cousin Francis Galton developed into the eugenics movement, with its monstrous legacy, and which came to a demonic conclusion in the Nazi Period with its ideology of biological racism.

One can see, therefore, in Darwin's own writings, a foreshadowing of the many issues which would make up the Darwinian debates, and their implications for educational philosophy. Is natural selection the only mechanism by which evolution moves forward, and must therefore all characteristics of all species be accounted for according to their contribution to individual fitness? Must all strategies for survival be based on self-interest, or can characteristics of kindness emerge out of the evolutionary process of natural selection? Are humans subject to the same laws of evolution, or does our intelligence and culture remove us from the evolutionary trajectory? Do, for example, differences between males and females throughout the natural world predict similar differences between men and women, or, by virtue of human plasticity, are differences between men and women primarily a result of culture, not biology? And finally, should natural law create cultural norms and ideals, or do we rebel against our natures, and create a society which stands opposed to the "ethics" of nature?

Darwin was interpreted in a multitude of ways partly because his work, as you can see, leaves itself open to such disparate interpretations. There is certainly room for those who want to see competition as the primary characteristic of the natural world, those who want to see such characteristics applied to human culture, and those who want to see Victorian England as the pinnacle of the evolutionary process. But there is reason to interpret Darwin differently. Darwin forcefully rejected transcendental views of human reason, religious or secular, as he placed human beings and rationality within an embodied idea of human existence. Although some interpreters of Darwin will continue to see human intelligence and culture as being fundamentally different from the rest of nature, Darwin placed human beings firmly within, and as part of, the world. And in describing that world, Darwin was adamant that cooperation is a possible strategy in nature, widespread among the social species. I believe that he presented a compelling argument that morality is rooted in our biology, understanding that culture expands on these intuitions, but does not invent them. And while clearly arguing that Victorian morality is a result of the evolutionary process moving ever forward, he simultaneously and powerfully rejected the upward trajectory of evolution. Darwin's chosen metaphor was a tree, and not a ladder, suggesting the nonlinear radiation of branches from a common trunk, with some continuing and leading to other branches, but others abruptly ending, essentially leading to the extinction of that branch. The success of one branch as opposed to another, as Darwin pointed out, is by chance, and not determined. A different branch could have just as easily survived in its stead.[71] Victorian England was not, according to this view, the end point of evolution, but just another option in the infinite variety of possibilities.

This is, obviously, not the only, or even dominant, interpretation of Darwin. The interpretive debate began already in Darwin's lifetime, launched, as argued, by Darwin himself. In the latter part of the nineteenth century, stretching until the 1920s, Darwinism was applied to a host of the human sciences—anthropology, economics, psychology, political science, sociology, and, of course, educational philosophy.[72] The more well known versions of Darwinism are, of course, the social Darwinist variety, but even there the reality is far more complex and sophisticated than the stereotyped version would have us believe.[73] Still, the idea of competition as the driver of progress does play the predominant role. I'm interested in the interpretive voice less noticed when discussing Darwinism, the one which I hold is both truer to Darwin's theory and, more importantly, truer to the reality of the human condition. In the next chapter, I look at the application of Darwinism to educational philosophy, focusing on the philosophical and educational ideas of Peter Kropotkin, the Russian prince and natural historian turned anarchist, and the first-generation debate about the meaning of Darwin for educational philosophy. Kropotkin's philosophical work is fascinating, although ultimately deeply flawed, suffering, like many of the first-generation Darwinists, including Darwin, from what I would call an unproblematic hermeneutic, as it moves freely from nature to culture, without recognizing the difficulty in applying nature's laws to the uniquely cultural landscape of humans. Still, Kropotkin's work is a first attempt to pay attention to the cooperative voice in Darwinism and to apply it, and to see what a Darwinian educational philosophy based on ideas of cooperation and community, rooted in the natural world, might look like. Later, I expand on Kropotkin's intuitions, looking at Dewey, who had a far more sophisticated hermeneutic and therefore sophisticated understanding of the implications of evolutionary thought for educational philosophy, and then, in our own time, at Mary Midgley, who I believe continues and significantly expands the tradition, although at times suffers from Kropotkin's error, as well.

Kropotkin's work is particularly interesting because he argues with the central advocates of two other interpretative streams of first-generation Darwinism—Herbert Spencer and Thomas Henry Huxley. Both Spencer and Huxley agreed on the competitive character of nature as its central characteristic. They differed in that Spencer viewed it as a characteristic which needs to be embraced, at least for a time, by human culture, while Huxley believes, like Dawkins later on, that human culture needs to rebel against its unseemly animal roots. Kropotkin's work, then, allows us a window into the first Darwinian debates, how Darwinism was applied to educational philosophy, and how a cooperative approach was articulated in response to the more famous Darwinian alternatives.

CHAPTER TWO

Nature's Lessons

Applying Evolutionary Theory to Educational Philosophy in the Nineteenth Century

Kropotkin is often seen as a historical oddity and, as a scientist, a curiosity. Born into Russian royalty, he rebelled against his privileged past, escaping from czarist Russia for exile eventually in England. He was an anarchist, and rooted his anarchism was his own particular interpretation of evolutionary theory. Kropotkin spent his early years as a naturalist, and, after spending several years in Siberia observing the natural world, wrote his own classic evolutionary treatise, *Mutual Aid*, which moves from cooperative claims about the workings of the natural world to anarchistic claims about human nature and society. Siberia was Kropotkin's "Beagle"; *Mutual Aid* his *On the Origin of Species* and *The Descent of Man*. Although both his biological and political treatises have been largely disregarded, many of his ideas, both biological and political, are still with us. And, as Stephen Jay Gould remarked in an essay on Kropotkin's evolutionary theories, he was no kook—his ideas are serious.[1] His modest educational writings were an extrapolation of his scientific, philosophical and political ideas, and, taken together, are all located within a larger debate between first-generation Darwinists on the implications of Darwinism for human society.

Kropotkin's *Mutual Aid* was written in response to an article by Darwin's "bulldog" Thomas Henry Huxley, in the British intellectual journal *The Nineteenth Century*, which Huxley later developed into his controversial Romanes Lectures of 1893, entitled "Evolution and Ethics."[2] And Huxley's work was a clear response to Herbert Spencer's earlier work.

To understand Kropotkin and the context in which he was writing, one should therefore begin with Huxley and with Spencer.

Huxley was a towering nineteenth-century scientific figure, credited for fighting Darwin's battles and playing a leading role in achieving evolution's acceptance within science and society. It is because of this that his claim, in his Romanes Lectures of 1893, that human nature stood in opposition to the natural process got such a reaction from other Darwinists—Kropotkin, Spencer, and, as we shall see later, Dewey, as well.[3] They felt that Huxley, by resurrecting the barrier between nature and human nature, had abandoned Darwin's most central philosophical idea.

Huxley described the evolutionary process in terms of the Social Darwinist stereotype of "nature red in tooth and claw." He described the essence of the struggle for existence as "self-assertion, the unscrupulous seizing upon all that can be grasped, the tenacious holding of all that can be kept."[4] In other words, selfishness, competition and radical individualism are the legacy of evolution. Human nature is a result and an exemplar of such evolutionary "values." The human success in becoming the dominant species that it has become is indebted to the competitive qualities which it shares with its relatives in the natural world. The human dominates as a result of " . . . his exceptional physical organization; his cunning, his sociability, his curiosity, and his imitativeness; his ruthless and ferocious destructiveness when his anger is roused by opposition."[5]

But humanity's emerging consciousness and ethical urgings are a break with the evolutionary "cosmic" process:

> But the influence of the cosmic process on the evolution of society is the greater the more rudimentary its civilization. Social progress means a checking of the cosmic process at every step and the substitution for it of another, which may be called the ethical process; the end of which is not the survival of those who may happen to be the fittest, in respect of the whole of the conditions which obtain, but of those who are ethically the best.[6]

Darwin, as we saw, already set the stage for such a view. He also saw the ethical and evolutionary processes as ultimately in tension with one another. Huxley developed the idea farther. Human beings, while emerging from the natural world, have evolved to be apart from it.[7] Human values are the antithesis of the values of the cosmic process. Where the cosmic process celebrates competition, the ethical process celebrates cooperation; where the cosmic process rewards radical individualism, the ethical process values community; where the cosmic process is driven by selfishness, the ethical process teaches selflessness.[8]

Huxley's dualism is a conscious critique of what he calls Herbert Spencer's "fanatical individualism."[9] Spencer was a leading advocate of what became known as social Darwinism. He was widely read and considered to be among the most important philosophers of his generation at the end of the nineteenth century, although his influence almost completely disappeared, as did that of Darwinism, by the 1920s.[10] For Spencer, competition among individuals of the species, in human beings as in the rest of nature, is the way in which evolution moves forward. Spencer saw evolution as marching ever upward. Only by allowing evolutionary forces to operate freely can progress occur. More than competition, then, and perhaps even more than individualism, Spencer believed in laissez faire principles. Society needs to operate according to its own natural laws, and any interference with them will obstruct the evolutionary march forward of civilization. Kropotkin is not the only Darwinist with anarchist sympathies. For Spencer, government, as a human-made institution, interferes with the natural development of society.

It should be mentioned that Spencer's social Darwinism was not always as the stereotyped view makes of it. Spencer's view of cultural evolution held that society moved forward incrementally, and that in each epoch the evolutionary strategies change. Now that we have reached what Spencer called the industrial epoch in society, cooperation becomes the chief mechanism for progress, as we emerge out of our primitive stage in which competition ruled.[11] However, while competition could give way to cooperation, progress was dependent on the individual being free to pursue self-interest (cooperation was an instrumental, not ethical, stance), without the hindrance of societal rules and regulations.

Huxley, while accepting the description of the natural world and at least early human civilization as selfish, competitive and individualistic, refused to allow Spencer's description to be seen as prescriptive for human society:

> As I have already urged, the practice of that which is ethically best—what we call goodness or virtue—involves a course of conduct which, in all respects, is opposed to that which leads to success in the cosmic struggle for existence. In place of ruthless self-assertion it demands self-restraint; in place of thrusting aside, or treading down, all competitors, it requires that the individual shall not merely respect, but shall help his fellows; its influence is directed, not so much to the survival of the fittest, as to the fitting of as many as possible to survive. It

repudiates the gladiatorial theory of existence. It demands that each man who enters into the enjoyment of the advantages of a polity shall be mindful of his debt to those who have laboriously constructed it; and shall take heed that no act of his weakens the fabric in which he has been permitted to live. Laws and moral precepts are directed to the end of curbing the cosmic process and reminding the individual of his duty to the community, to the protection and influence of which he owes, if not existence itself, at least the life of something better than a brutal savage.[12]

What allows human beings to free themselves from the fate of the cosmic process? For Huxley, it is reason which allows human culture to develop an alternative "human nature," in contrast to evolutionary "nature."[13] Human reason can conquer the emotions, the signatory characteristic of the species.[14] Human behavior which is solely motivated by human emotions, while "natural" in one sense, is contrary to the higher calling of human "nature," and therefore can be considered "pathological."[15] The human being is in a psychological struggle between his/her given evolutionary nature, and his/her higher nature. Societal progress is dependent on the higher nature of reason winning out over the lower nature of emotions. The Stoics, Huxley pointed out, often called this higher nature of reason "political" nature.[16] For the Stoics, human beings "living according to nature" meant living a rational life.

As has been pointed out by Paradis, Huxley's psychology strongly foreshadows that of Freud.[17] The human being is divided into two selves, the natural self of evolution, represented by the passions/emotions, and the natural self of society, represented by reason. Huxley, like Freud after him, argued that the two sides of the self are in perpetual conflict, and that while reason can control the emotions, it cannot conquer them. The emotions need to be appeased, but cannot be let loose, for they are at odds with the needs of society. Like Freud's psychology, Huxley's is a pessimistic one. There is no possibility of resolving the struggle, but only in constantly and vigilantly managing it, a point which, as I have pointed out, Dawkins also advocates in our day.[18]

Huxley used the metaphor of the gardener to argue his point.[19] The gardener fences in nature and cultivates it. The garden is a work of beauty—but it is an artifact, a work of human beings. Although emerging out of nature, its beauty is dependent on fencing it off from nature, and protecting it against noxious weeds and pests which come to destroy it. The gardener, therefore, must protect the garden from nature. Nature's aims are in direct conflict with the aims of the gardener.[20] This is a metaphor which Dewey also used, largely reversing its meaning and through it, challenging Huxley's dualism, as we shall see.

While wandering the harsh and marginal ecosystems of Siberia, Kropotkin came to believe that Darwin, or at least Darwinists, had gotten the description of nature and its evolutionary mechanisms wrong. As Gould pointed out, the character of the ecological landscape of Siberia had a particular influence on Kropotkin's environmental worldview, as did the tropics on Darwin's.[21] Kropotkin accepted that competition was the main mechanism of natural selection. However, citing Darwin, Kropotkin argued that competition was largely metaphoric, and referred to a number of strategies.[22] Darwin had listed three different competitive strategies which fell under his metaphoric idea of "struggle for existence." The first, competition between individuals, was the one that Huxley had presented as the sole meaning of struggle. Darwin held this to be the primary kind of competition, although Kropotkin contended that Darwin changed his mind about this over the years.[23] Kropotkin, however, saw Darwin's second and third meanings of "struggle for existence" as the prominent ones in nature: competition between groups, what is now known as "group," rather than "individual" selection, and competition between the species and the environment.[24] As Darwin himself pointed out, competition between the species and the environment is most prevalent in extreme climates, where the struggle for existence is not between individuals for limited resources, but rather the struggle to develop strategies for exploiting a seemingly inhospitable landscape.[25] Siberia was an archetypical example of such a place. Kropotkin therefore, refused to accept that competitive *behavior* between individuals was the most prominent and effective strategy for ensuring survival within the natural world.[26] Both of the other meanings of "struggle for existence," group versus group, or group versus the environment, led to cooperative behavior between individuals, rather than competitive behaviors.[27] Nature was not "red in tooth and claw" at its most basic level, but rather nurtured cooperation, chiefly within species.[28] For Kropotkin, mutual aid, one possible "competitive" strategy in nature, is the predominant rule of natural selection.[29] Kropotkin's understanding of the origins of mutual aid is similar to Darwin's description of the evolution of the moral sentiments, but for him the instinct of community, sociability and mutual aid seem to be shared by a far larger sampling of species than Darwin maintained.[30] Sociability and mutual aid are the most effective strategies for species' survival, and progress:

> The animal species, in which individual struggle has been reduced to its narrowest limits, and the practice of mutual aid has attained the greatest development, are invariably the most numerous, the most prosperous, and the most open to further progress.... The unsociable species, on the contrary, are doomed to decay.[31]

Such a view of nature clearly made it easier to posit a normative role for the laws of nature. As Kropotkin stated, when nature is perceived as "an incessant struggle for life and an extermination of the weak ones by the strongest, the swiftest, and the cunningest, evil was the only lesson which man could get from Nature."[32] Kropotkin's assumption was that such a competitive view of nature would support a separation between the descriptive and the prescriptive, suggesting that ethics have an "extra-natural" or "supernatural" source, since nature's selfish, individual, competitive character cannot possibly serve as an origin for an ethic of altruism, community and cooperation.[33] As Kropotkin argued, it is not surprising that Huxley, having identified nature with individual competition, needed to look elsewhere for morality, as is discussed shortly.[34]

Kropotkin's view of the natural world, however, fit perfectly with his moral worldview. While Huxley and others would eventually protect accepted notions of morality by re-erecting the barrier between the natural and the moral, Kropotkin defended his notion of the moral through his view of the natural. The concept of mutual aid was not simply making a scientific argument about the nature of the world, but a moral argument that nature can and should be trusted.[35] Kropotkin believed that nature and its processes serve as the conceptual basis for human morality. Since nature is at its root cooperative, civilization's competitive and warlike character was a result of history, not biological determinism. Civilization has distanced itself from its true instinct, and instead accepted this competitive, aggressive, violent view of human nature, which is mistaken but self-actualizing.[36]

Kropotkin, for example, argued that the Eskimos originally held a communist concept of property, which later degenerated due to Western influences: "Eskimo life is based upon communism. What is obtained by hunting and fishing belongs to the clan. But in several tribes, especially in the West, under the influence of the Danes, private property penetrates into their institutions."[37] Kropotkin argued that mutual aid is part of our evolutionary legacy, but that competition has become culturally dominant because of societal choices. It is an argument that has deep roots in the Romantic tradition, and was heard often in the early anthropological environmental literature, as it attempted to document that the natural state of indigenous peoples was socially cooperative, economically communal and environmentally sustainable, and that the "fall from Eden" came as a result of Western colonialization.[38]

In spite of the power of socialization, however, Kropotkin did not believe that it could eliminate our innate human nature. Our natures remain as a resource to resist socialization when it is dehumanizing, that is, when it is antithetical to our natures:

> In short, neither the crushing powers of the centralized State nor the teachings of mutual hatred and pitiless struggle which came, adorned with the attributes of science, from obliging philosophers and sociologists, could weed out the feeling of human solidarity, deeply lodged in men's understanding and heart, because it has been nurtured by all our preceding evolution.[39]

The existence of a precultural layer of human nature existing independently of socialization is, of course, a controversial notion. It is an idea that is presented by Dewey, as well, and developed by Midgley, as we shall see.

Kropotkin did not believe that mutual aid is identical with morality. Nature selects for sociability, not morality. Like Darwin, who believed that there is a social instinct which is the basis out of which morality develops and interacts, Kropotkin also believed that sociability coevolves with intelligence, and that intelligence then expands mutual aid into morality.[40] Kropotkin viewed mutual aid as an evolutionary strategy of survival which, while allowing individuals to more effectively survive and replicate through cooperation with a group, nonetheless can be selfishly motivated or encourage competition between groups. Mutual aid, for example, flourishes because nature gives advantage to those who recognize that their own survival is enhanced if they cooperate with others who return the favor; or, that survival is enhanced when the group defends itself as a group, or attacks other groups. In this case, mutual aid, while encouraging cooperation within the group, simultaneously encourages aggression between groups.[41]

The idea that we are hard-wired to be loyal to our own as the flip side of being suspicious and hostile to "the other" is an idea to which Darwinian political scientists are beginning to pay attention, although it is also resisted for being reductionist and simplistic in its understanding of the nature–culture interaction.[42] Kropotkin first draws our attention to the possibility that our affections for our own comes with the price biologically of suspicion for others, while also believing that our intelligence can breach the limitations of mutual aid, and extend sympathy beyond its biological origins. Intelligence guided by the social instinct can trump the innate limits of mutual aid.

Kropotkin contended that true morality is altruistic, identifying the well-being of others so completely with one's own well-being that one does not expect any benefit in return for one's actions.[43] Justice transports us from mutual aid to true morality, expanding the circle of mutual aid and cooperation to all of humanity. Kropotkin believed that the development of mutual aid as an evolutionary strategy of survival is linked to the development of intelligence, and therefore simultaneously leads to the development of the principle of justice in human communities, which is the application of the principle of mutual aid to the whole species. Breaking down the barriers between groups, recognizing that we are all part of one struggle for existence, is an intellectual leap which mutual aid might suggest, but only rationality can realize. True morality is the logical extension of justice, as our innate sympathy combines with our intelligence to break down the barriers between our own egos and the well-being of others, creating a world community where true altruistic cooperation flourishes.[44] This is not as strange an idea as it might first appear. Continuing the Aristotelian tradition, it sees our values embedded in our biology, and not transcendent of them. While the application of nature's laws to human actions might seem crude, the idea that our intelligence expands on a set of intuitions, rather than creating them ex nihilo, is a sound one. In subsequent chapters, a more sophisticated presentation of similar sentiments is discussed.

Kropotkin did not only attack the idea of competition. For him, this competitive view of human nature was connected to a view of the individual as radically independent and unconnected. While Spencer indeed idealized the radically independent individual associated with Social Darwinism, Kropotkin strongly rejected it. Communities of individuals built on mutual aid and even self-sacrifice, rather than radically independent individuals functioning on self-interest, are both the real and the ideal.[45] Mutual aid is both about the cooperative nature of human beings, and about their identities as parts of a larger whole. While for Spencer the State is evil because it forces the artificial claims of the society on the freedom of the individual, for Kropotkin the State is wrongly conceived because it is based on assumptions of the competitive individual: "a loose aggregate of individuals" whose connection to one another is purely instrumental.[46] Kropotkin's ideal is not a social contract which defines what is just, but rather a community in which the individual freely gives for the good of the whole:

> The higher conception of "no revenge for wrongs," and of freely giving more than one expects to receive from his neighbours, is proclaimed as being the real principle of morality—a principle superior to mere equivalence, equity, or justice, and more conducive to happiness. And man is appealed to be guided in his acts, not merely by love, which is always personal, or at the best tribal, but by the perception of his oneness with each human being.[47]

A society envisioned by the implications of mutual aid will be one with "an infinity of associations" which will "tend to embrace all aspects of life."[48] Life's meaning, as learned through evolution, is not the atomized pursuit of individual survival and material prosperity, but rather is reaching out into the world and recognizing our connection to others through our innate sociability.

Yet, in spite of his critique of radical individuality, Kropotkin was aware of the potential conservative nature of community and socialization, and therefore valued individuality as a counterbalance to its dangers, a place where innate nature can still reside independently of cultural norms. The "self-assertion of the individual . . . breaks through the bonds, always prone to become crystallized, which the tribe, the village, community, the city, and the State impose upon the individual."[49] In a later stage of development, the individual and the group can become one, as the individual recognizes that his/her individual fulfillment is dependent on social interdependence. But programmatically Kropotkin recognized a dialectical tension between the individual and the group.[50] The double tendency of what Kropotkin called sociality and intensity of life are eventually intertwined, so that life's meaning is found within sociality, and not apart from it.[51]

FROM PHILOSOPHY OF NATURE TO EDUCATIONAL PHILOSOPHY

Huxley's, Spencer's and Kropotkin's philosophies of nature set the stage for their three approaches to educational philosophy. They are, in fact, excellent case studies in the translation of philosophy into educational principles, and the many ways in which Darwinism was applied to education. While Spencer articulated what can be called a social Darwinist educational philosophy, Huxley ultimately retreated to a dualist model of nature and human beings. Kropotkin tried to navigate a third way: as opposed to Huxley, rooted in the natural world; but as opposed to Spencer, where the natural is defined by cooperation, not competition. Understanding how

Spencer and Huxley viewed education once again creates the context for understanding what Kropotkin was trying to do, and why.

SPENCER'S EDUCATIONAL PHILOSOPOHY

Spencer's ideas of evolution informed his ideas of education. And since his ideas of evolution have been so thoroughly disregarded, one would suspect that his educational ideas should follow a similar path. But as Kiergan Egan has noted, Spencer's ideas are in fact very much still with us.[52] While many of his writings are properly understood as reactionary, in a history of progressive education Cremin argues that if the progressive revolution had a beginning, it was surely the work of Herbert Spencer.[53] These conflicting understandings about his legacy are part of a larger discussion on the history of Darwinism. Branded already by the beginning of the twentieth century as reactionary, scholars have only in the last several decades broken out of such a stereotyped view of Darwinism and begun to define it as a movement with a much more complex and often conflicting social and political agenda.[54] Even the more reactionary interpretations of Darwinism, as popularly understood under the category of social Darwinism, turn out to be more nuanced than the stereotyped version gives credit for.

The key to understanding Spencer's progressive educational tendencies is to be found in the liberating role which nature takes in his philosophy. According to Spencer, evolution moves inexorably forward as each epoch progresses into the next. Cultural evolution is dependent on the same laws of nature. Because Spencer's views are in fact more rooted in Lamarck's theory of evolution than in Darwin's theory of natural selection, societal progress is dependent on each generation "stretching its neck" slightly farther, so that the next generation begins at a more advanced stage than the former.[55] For Spencer, the educational goal is to facilitate the incremental stretching of the neck. Here is the central tension between Spencer's reactionary and progressive tendencies: radical change cannot take place—change must occur at nature's pace—but simply accepting the norms of the epoch is also not an option.

The result is an education which simultaneously educates us for life in this epoch, but also for life in the epoch which we are ushering in. When focusing solely on an education which trains us to live within the constraints of contemporary society, his educational views sound conservative and conformist:

> Is it not that education of whatever kind has for its proximate end to prepare a child for the business of life—to produce a citizen who, while he

is well conducted, is also able to make his way in the world? And does not making his way in the world (by which we mean, not the acquirement of wealth, but of the funds requisite for bringing up a family)—does not this imply a certain fitness for the world as it now is?[56]

But when focusing on educating for the coming epoch, one can hear the progressive perspective, with its biases, emerging:

> On the other hand, education, properly so called, is closely associated with change—is its pioneer—is the never-sleeping agent of revolution—is always fitting men for higher things, and *unfitting* them for things as they are. Therefore, between institutions whose very existence depends upon man continuing what he is, and true education, which is one of the instruments for making him something other than he is, there must always be enmity.[57]

Such an education, not surprisingly, was at odds with the view of education of the day. For Spencer, Victorian education taught the "ornamental" rather than the "useful"; "appearance" rather than the "functional"; the "conventional" rather than the "intrinsic"; "what we shall be" rather than "what we are."[58] Rather than giving tools for complete living according to the "true" demands of life as dictated by natural evolutionary laws, education taught socialization into a culture which was far removed from living according to the laws of nature. Random historical facts, extinct languages and ancient superstitions were all part of a classical education, whose focus, in Spencer's view, was conformity to the world around them, rather than obtaining skills for living life according to the laws of nature, and therefore facilitating societal evolutionary progress.[59] Spencer's educational ideal was not primarily adaptation to culture, but rather adaptation to the world of nature, as defined by evolution. Tradition and society do not define educational goals; nature does.

How was socialization to be avoided? Through a radical view of the individual freeing him/herself from the constraints and expectations of society. Indeed, Spencer's evolutionary view has at its center a laissez faire individualism which resists the demand to impress or to conform, to suppress or to be suppressed. His educated individual follows the laws of nature which stand independent of the authority of the cultural past.[60] "Survival of the fittest" means, foremost, the survival of the fittest individuals, those individuals who most conform to the evolutionary laws of nature, and not the laws of (Victorian) culture. The advancement of culture is completely dependent on the individuals within society adapting themselves to natural law.⁶¹ For Spencer, those who lived closest to nature's laws would be those who were the fittest to survive, and consequently those who facilitate the progress of society as a whole. When focusing on proximate ends for education, one might think that education teaches conformity to societal norms. When focusing on ultimate ends for education, however, it becomes clear that it is nature which needs to be conformed to (ultimate ends), not society (proximate ends). The focus on the individual as an independent unit frees him/her from the constraints of conformity, and allows the independence necessary for evolutionary progress.

How are these natural laws to be accessed? Spencer gave two seemingly contradictory answers. On the one hand, Spencer argued that the instincts, emotions, feelings and sensations are, in principle, trustworthy guides to nature's laws.⁶² On the other hand, however, Spencer celebrated science and its rationality, which through its access to the truth of human nature and the nature of the world, solves the riddle of nature's laws and, in turn, how one is to live a full life. Nature's laws can be accessed not only through authentic intuitions, but also through rational, scientific thought.

Why are intuitions to be trusted? As products of evolution, humans are part of the natural world. As such, both human nature and the human relationship to the natural world are expressions of the laws of nature. Human instincts therefore are, in principle, trustworthy indicators of proper human behaviors and attitudes:

> Happily, that all-important part of education which goes to secure direct self-preservation, is in great part already provided for. Too momentous to be left to our blundering, Nature takes it into her own hands. While yet in its nurse's arms, the infant, by hiding its face and crying at the sight of a stranger, shows the dawning instinct to attain safety by flying from that which is unknown and may be dangerous. . . . ⁶³

Instincts are nature's laws manifested in behavior. Human emotions are an accurate compass for navigating human physical, intellectual, and moral development. Because of the centrality of the instincts, emotions, feelings and sensations to Spencer's worldview, it is not surprising that he sees them as the predominant component of human nature. Rather than intellect, it is the world of feeling that forms the foundation of human nature, as it does for other animals. As Spencer put it, "the chief component of mind is feeling."⁶⁴

Yet, according to Spencer, we can no longer trust our instincts for knowing the correct way to behave. Socialization to the wrong laws has made our feelings untrustworthy. Since Victorian education teaches conformity to a society which emphasized the ornamental rather than the functional, individuals have been taught to distrust and distance themselves from the natural (and their feelings), focusing on brain and not body, thought and not emotion:

> It is true that, in those who have long led unhealthy lives, the *sensations* are not trustworthy guides. People who have for years been almost constantly in-doors, who have exercised their brains very much and their bodies scarcely at all, who in eating have obeyed their clocks without consulting their stomachs, may very likely be misled by their vitiated *feelings*. But their abnormal state is itself the result of transgressing their feelings. Had they from childhood never disobeyed what we may term the physical conscience, it would not have been seared, but would have remained a faithful monitor.[65]

Nature's laws are imprinted in human nature, but culture has been able to undermine its authority. Living according to nature, humans would grow up to lead healthy, moral lives, but society has distorted the clarity of nature's voice.

Science is the other interpreter of natural law. It is critical to the future of evolutionary progress since it allows us to rediscover what natural law is, and return to the correct path.[66] Science, as the only avenue remaining to discover the laws of nature that are necessary for evolutionary progress, is crucial to human self-understanding. The more parents, teachers, and politicians understand these laws, the more it will be possible to rediscover how to live according to them, and therefore, how one is to live.[67]

Science education stands as the antithesis to classical education. It is not simply another subject for the curriculum, but rather the heart and soul of a Spencerian education. Freed from the shackles of irrelevant traditional subjects, science teaches us what it means to be human, and how best to act in the world. Echoing Darwin's comment on studying baboons rather than studying Locke in order to learn metaphysics, Spencer argued that the study of the science of ethology is far more worthwhile to human life than the study of the classics.[68]

Which path is more trustworthy—that of instinctual feelings or that of scientific reason? For Spencer, the path of feelings has lost its authority due to the decaying influence of culture. Culture has confused the individual's ability to understand him/herself.[69] Science, in this case, provides an avenue, outside of the confusion of culture, in which the true laws of nature can be clarified. The rational has access to how the emotions function, and can return the emotions to the proper path. Intellectual knowledge can lead to a return to a healthy emotional life. Rationality and feelings are not in tension; intellectual understanding allows us to identify our inner natural voice and respond correctly to the world around us. Once able to identify our correct inner voice, we can once again trust our instincts to guide us. Our science and our instincts will once again be in harmony.

Curriculum: What to Teach?

What follows from all this is that there are two defining foundations to Spencer's. The first is the authentic voice of feelings, emotions and instincts. The second, of course, is science. They are interconnected.[70]

Individuals need to hear their authentic inner voices, without the disruption of the often corrupting influence of society. This foundation of Spencer's curriculum led him to identify with the reformist movement of the nineteenth century. Spencer identified his views with those of Pestalozzi, who continued the educational tradition of Rousseau.[71] Children, rather than being socialized into the larger society, need to be able to develop according to their own unique individualities. It is the individual, allowed to develop according to his/her own nature, who contributes to societal development. As with the Romantics, for Spencer human nature is foremost identified with the life of the emotions, which are the "masters," and the intellect are their "servants."[72] It is the robust development of one's nature as expressed in one's feelings, emotions and instincts that is central to individual, and therefore societal, progress. Psychological insights for adapting educational curricula to stages of child development are part of such an approach. The curriculum needs to be age appropriate and adapted to the child's abilities and interests at all times. It is the child's natural interests and curiosities which drive the curriculum, and not society's mistaken view of what a child needs to know, or what a child should be. The goal of the curriculum is the formation of an adult of character, an adult who is fully true to his/her nature.[73]

For the same reason that Spencer emphasized feelings, emotions and instincts over rationality, he also emphasized body over mind. He reasserted the importance of the emotional and the physical as essential components of human life, and therefore central focii of education. He rejected the overemphasis in education on mind over body, and therefore saw physical education, for example, as a central component of his curriculum.[74] More than a philosophy of the body as a means to an end, in Spencer's view the development of a fit body is at least as central to being fully human as the development of reason and rationality. Survival of the fittest is still, at least in our epoch, about physical fitness.

However Spencer, as has been shown, rejected relying solely on emotions/feelings/instincts to access nature's laws, and, therefore, to build an educational philosophy, and in turn an educational curriculum, for two reasons. The individual's ability to hear his/her authentic voice has been damaged by society. As noted, the individual has become confused, no longer able to distinguish between the "truth" of the authentic individual, and the superimposed "truths" of society. In addition, the gap between a human nature of proximate ends and a human nature of ultimate ends suggests that the individual cannot rely on his/her instincts in order to progress. A mechanism is necessary which allows the individual to move beyond his/her current nature, and to evolve a more advanced one.

That mechanism is rationality, and science is its result. Science is the key to understanding human nature, and critical at a time when the other source of identifying authentic human nature, emotions/feelings/instincts, has been distorted. Furthermore, science not only can describe and identify proximate human nature, but by understanding the underlying evolutionary processes it can also identify ultimate human nature, and with it the utopian ends of societal evolution. It is, therefore, a compass for societal progress. "What knowledge is of most worth?" Spencer asked rhetorically. "The uniform reply is—Science."[75]

The positivist movement of the nineteenth century saw science as objective truth, and as the means to societal progress. For Spencer, science was in fact the only knowledge of real worth.[76] The uncovering of the laws of nature allowed us to understand not only who we are, but since nature is progressing, allows us to understand the mechanisms of progress. This, in turn, determines proper actions, those actions which allow nature to evolve without the interference of society. Science provides knowledge untainted by society and societal norms and mores. Scientific method and knowledge is, therefore, the other critical component of the educational curriculum. Spencer's definition of science is, however, extremely broad, and it includes the social, as well as the natural, sciences. For example, since society is a natural, organic phenomenon, it can be only understood through nature's laws.[77] Spencer is one of the founding fathers of sociology, and is considered one of the key social scientists of the nineteenth century, the term itself memorializing a positivist notion that science can explain the social world as well as the natural one.⁷⁸ Aesthetics can also be explained through nature's laws. Which music and art are of intrinsic worth, for example, are determined by that which adheres most closely to the scientific laws of nature. Good music emerges from the proper use of the language of the emotions, as understood by science; good art from the proper use of perspective and color. The good, once again, is synonymous with the true expression of the laws of nature.⁷⁹

In Spencer's view, philosophy is the ultimate science because philosophy integrates the laws of science, and therefore teaches us how to live. Philosophy is a science because it is not simply about teaching facts about the laws of nature, but rather integrating nature's laws into an understanding of how the world works, and therefore how humans work and why. The curriculum, anchored in science, has failed if it has taught random facts which the student cannot integrate into prescriptive action. At its best science teaches how to compose and appreciate music, how to paint a picture, how to make a living, and, of course, how to live a moral life. Science's worth comes from its prescriptive, and not only its descriptive, value. Philosophy is the ultimate science which ensures integrative thinking and, as a result, right living.⁸⁰

Pedagogy: How to Teach?

The same ideas of a "natural" education are applied to pedagogy. Learning is not something that needs to be imposed on the child by society. Curiosity and learning are instinctual and natural to human beings, particularly children:

> Every botanist who has had children with him in the woods and lanes must have noticed how eagerly they joined in his pursuits, how keenly they searched out plants for him, how intently they watched while he examined them, how they overwhelmed him with questions.⁸¹

If learning is natural, the primary role of educators is to let nature take its course: "What we are chiefly called upon to see, is, that there shall be free scope for gaining this experience and receiving this discipline—that there shall be no such thwarting of Nature."⁸² Nature's pedagogy needs to be allowed to work with its students, and educators need to facilitate this interaction. In other words, educators need to get out of the way.⁸³ Like government, educational interventions are more likely to derail the natural process than assist it. Spencer's anti-State liberalism expresses itself pedagogically in a hands-off approach to education.

In much the same way that children naturally learn, Spencer believed that the natural consequences of children's actions would be their reward or punishment. Lazy children would be punished in life for their laziness; mean children will be related to meanly by others; cowardly children would not succeed in a world where the fittest survive.[84] For Spencer, punishment naturally emerges from flawed deeds.[85] It is not imposed by society, but is rather a natural outgrowth of maladaptation, acting as a natural tool for the education of the young in the laws of nature/society. Punishment, when instilled, should be a clear and obvious consequence of the inappropriate behavior, what Spencer coins a "natural discipline."[86] Spencer's pedagogy, like his entire philosophy, blurs the distinction between "is" and "ought." The natural is prescriptive, not simply descriptive.

Although Spencer speaks at times of teachers, and toys at one point with the idea of setting up his own school, no less often Spencer speaks of parents as the central educational figures.[87] Since parents are from an evolutionary perspective the natural educators, this should not be surprising. Teaching the young is a central activity in many species, particularly the human species. Spencer points out that the extremely long period of dependence of human children on their parents indicates the central role evolution gives to the education of the human species.[88] The fact that parents have abdicated their natural roles and have begun to pass on responsibility to the State, as mandatory schooling began to be debated and eventually implemented in late nineteenth-century England, is a dangerous trend that disrupts the natural basis of education.[89] The home schooling movement might have found an enthusiastic advocate in Spencer.

Spencer's pedagogy is based on fostering independence in the child. As such, he rejects a pedagogy which makes the student a passive recipient of information or rules. Rote memorization, a standard part of Victorian pedagogy, was seen as stifling the child's innate curiosity, resulting in passive, submissive adults. The teaching of language, for example, teaches passivity as children are taught that knowledge needs to be received, and not discovered: "The learning of languages tends, if anything, further to increase the already undue respect for authority. Such and such are the meanings of these words, says the teacher or the dictionary. So and so is the rule in this case, says the grammar."[90]

Education should foster independent learners. Just as science stood as the cornerstone of Spencer's curriculum, so too science stood as the cornerstone of Spencer's didactics. Science, in this case, is not a body of knowledge for the curriculum, but a pedagogy, a methodology of how to learn. Science teaches through experience and the discovery of knowledge, rather than the passive acquisition of information: " . . . in education the process of self-development should be encouraged to the uttermost. . . . They should be *told* as little as possible, and induced to *discover* as much as possible."[91]

Education should be something that is natural to the child, and therefore something which is enjoyable.[92] Spencer envisioned children running outside, turning over rocks, examining insects, being enthralled with the joy of learning. He saw educators (primarily parents) as enthusiastic learning partners, not teaching by the power of authority, but through friendship and example.[93]

Spencer's pedagogy, a direct result of his educational philosophy, explains clearly why he was seen as a forefather for the progressive educational movement. In the move from the teacher's authority to the child's independence, from passive memorization to active discovery, from discipline to freedom, all justified by the implications of evolutionary theory as he understood it, Spencer was advocating a radical change in educational practice, equally for girls and for boys, it should be noted.[94] This, of course, should not eclipse his conservative, often reactionary views. While both sides of Spencer's ideas are deeply problematic, the intellectual roots of his progressive tendencies are what still have influence, and can be examined to explore their relevance.

Spencer's educational goal, of course, was the fostering not only of free and independent individuals, but of a society and culture rooted in such values. Spencer believed that a pedagogy of discovery and self-motivated learning taught such values, and would ultimately foster societal progress. Order and discipline at home or in the classroom breeds passivity. A pedagogy of noise, activity, freedom and independence would allow a truly free society to evolve:

> The independent English boy is the father of the independent English man; and you cannot have the last without the first. German teachers say that they had rather manage a dozen German boys than one English one. Shall we, therefore, wish that our boys had the manageableness of German ones, and with it the submissiveness and political serfdom of adult Germans? Or shall we not rather tolerate in our boys those feelings which make them free men, and modify our methods accordingly?[95]

HUXLEY'S EDUCATIONAL PHILOSOPHY

Given Huxley's fierce opposition to Spencer's views on human nature, it follows that his educational philosophy would move in a very different direction. At first glance, however, it might not appear so. Huxley's educational writings come from a period predating his dualist position explicitly articulated in the Romanes Lectures. While present even in the early Huxley, the dualism is not nearly as pronounced.[96] Scholars have disagreed as to whether there is a shift in Huxley's position over the years, from a romantic to a dualist one, or whether the dualism was always there. I think it is fair to argue that a difference in degree developed into a difference in kind. In Huxley's educational writings, the dualism is already clearly visible, yet with noticeable romantic tendencies which seem to contradict his dualist position. In other words, Huxley always maintained a distinction between what he later called "the cosmic process" and "the moral process." However, where in his earlier writings such a distinction is interspersed with romantic notions of being at one with nature and learning from nature's laws, in *Evolution and Ethics* the two categories became clearly antithetical. One can see in Huxley's educational writings many of the elements one would expect from his later dualist philosophy, although still couched in a language of continuity with the evolutionary story.

Like all the first-generation Darwinists, Huxley contends that education is about learning to live according to the laws of nature. Nature's laws define the human *telos*. It is what gives human life its meaning. Education is not only about material success, it is also about human happiness by virtue of living life the way that it was meant to be lived.[97] Huxley defines this as the goal of a true liberal education, for which he was a leading advocate in the second half of the nineteenth century.[98]

Nature, in Huxley's educational writings, is not morality's enemy, as Huxley later argues in *Evolution and Ethics*, but is rather, like a romantic might argue, the guiding principle for human action.[99] Morality, therefore, in this earlier conception, is learned from the natural world:

> Let us consider what a child thus "educated" knows, and what it does not know. Begin with the most important topic of all—morality, as the guide of conduct. The child knows well enough that some acts meet with approbation and some with disapprobation. But it has never heard that there lies in the nature of things a reason for every moral law, as cogent and as well defined as that which underlies every physical law; that stealing and lying are just as certain to be followed by evil consequences as putting your hand in the fire, or jumping out of a garret window.[100]

Huxley, like Spencer, is arguing that punishment for immoral acts will be delivered, not only by human society, but by nature itself. This is quite different from his later position on morals, in which nature in fact rewards immoral acts, and society needs to punish those immoral acts which nature rewards, and to encourage behaviors which nature punishes. Here, as with Spencer, morality is part of nature's laws, and immoral acts, by disobeying the natural order, are punished through the natural order.[101]

Morality, then, is a product of natural laws. The early Huxley argues that morality is learned from the natural (cosmic) process. Like many of the Darwinists, including Darwin, Huxley identified with Hume's view of morality, with morality's beginnings in the intuitive, emotional life of human nature, and devoted an entire work to his exploration of Hume's psychology and philosophy.[102] Huxley's commitment to natural law governing human life had implications for how he viewed the process of learning. Like all first-generation Darwinists, Huxley also believed that human beings were not born as a lump of clay to be molded by society, but rather that they were born with instincts which reach out into the world, and which seek interaction and communication. Language is a result of the natural desire to interact, not its cause, an idea which Dewey developed and which today resonates with other, particularly Darwinian views of language acquisition.[103] Learning is, therefore, a natural process—an instinct which is present in human nature. Learning also takes place within the natural world, in interaction between the inquisitive human being and the stimuli of the world to which s/he adapts. As with Spencer and, as shall be discussed, Kropotkin, also for Huxley the natural world itself is the prime teacher, and nature's lessons are mandatory for living. Even without schools, learning will take place; it is a question of survival:

> To every one of us the world was once as fresh and new as to Adam. And then, long before we were susceptible of any other mode of instruction, nature took us in hand, and every minute of waking life brought its educational influence, shaping our actions into rough accordance with nature's laws, so that we might not be ended untimely by too gross disobedience. Nor should I speak of this process of education as past, for any one, be he as old as he may. For every man the world is as fresh as it was at the first day, and as full of untold novelties for him who has the eyes to see them. And nature is still continuing her patient education of us in that great university, the universe, of which we are all members—nature having no Test-Acts.[104]

Continuing the progressive educational themes of Spencer, Huxley contends that education needs to mimic the natural interaction between the human being and the world. Pedagogy can't be based on passively sitting in the classroom collecting facts, but needs to foster active interaction with the world. Education needs to apply all the human senses, evolved for learning, to the process of learning.[105] This empirical exploration of the world is the most important lesson of education. As Huxley argues, presenting the progressives' case against the authority of tradition, "And, especially, tell him [the student] that it is his duty to doubt until he is compelled, by the absolute authority of Nature, to believe that which is written in books."[106] In nature, therefore, are the meaning and lessons of life, not in the classroom, or textbook, or teacher. The romantic assumptions of nature as a purer, truer, more accurate conveyor of truth and values are certainly part of his worldview.

"Nature's education," therefore, defines both the aims of education and its pedagogy. Nevertheless, and here the continuity between Huxley's earlier and later works can be discerned, nature's education is neither foolproof nor seamless. Huxley, even in his early work, argues that there is a need for "artificial education" in order to rectify "the defects in nature's methods."[107] His use of the term "artificial" is not arbitrary. The rules of life in human society are different than the rules in the natural world, and one taught only in nature's ways is ill-prepared for life in a cultural world.

Although Huxley's artificial education is presented as continuous with the natural world, it also has the seeds of being at odds with it. Unlike natural education, artificial education has its core in rational thought, as opposed to the "natural" tendencies of emotions and instincts.[108] Artificial education can be described as the evaluating of the "natural" by the "artificial," and the adapting of the "natural" into culture.[109] Although the artificial grows out of the natural, it begins to supersede it. Huxley's later distinction between "cosmic nature" and "political nature" is foreshadowed here. Rationality is the prime agent of human nature, and of an "artificial education." It is the filter through which the messages and images of a natural education are to be evaluated:

> That man, I think, has had a liberal education who has been so trained in youth that his body is the ready servant of his will, and does with ease and pleasure all the work that, as a mechanism, it is capable of; whose intellect is a clear, cold, logic engine, with all its parts of equal strength, and in smooth working order; ready, like a steam engine, to be turned to any kind of work, and spin the gossamers as well as forge the anchors of the mind; whose mind is stored with a knowledge of the great and fundamental truths of nature and of the laws of her operations; one who, no stunted ascetic, is full of life and fire, but whose passions are trained to come to heel by a vigorous will,

the servant of a tender conscience; who has learned to love all beauty, whether of nature or of art, to hate all vileness, and to respect others as himself.[110]

In Huxley's liberal education, therefore, the curriculum focuses first and foremost on science, and no less importantly on scientific methods. Unlike Spencer and, as we shall see, Kropotkin, where the scientific curriculum stands in dynamic tension with developing the "emotional" side of human nature, Huxley's view of the development of human nature is synonymous with his view of the development of human rationality. Because reason is the most characteristic feature of human nature, science is the most important discipline for the school curriculum, since it teaches rationality, while also teaching the laws of the cosmic process, which are critical for putting nature to proper and efficient use.[111] The curriculum is based on an artificial education, of which science is its most important component. Huxley never alludes to a complimentary curriculum of a natural education, which particularly for Spencer was symbolized by allowing human natural instincts to develop. Like the Stoics, Huxley's "natural" instinct is reason.[112]

Science—both its method and its findings—sits at the heart of Huxley's educational philosophy. Most important to his view, and this he shares with most of the Darwinist educational philosophers, is a pedagogy which rejects the authority of tradition and places independent rational thought as exemplified by science in its stead. Science here stands not only as a body of knowledge about how the world works, but as a methodology about how truth is to be discovered. No longer the rote memorization of random facts, but rather the development of skills with which to discover, evaluate and critique knowledge. Not the learning of mathematical formulas and equations, but the learning of the mathematical method.[113] The success of evolutionary theory showed the power of the scientific method. The triumph of the evolutionary story was a triumph of rationality over superstition, truth over tradition, and free inquiry over authority. The educational revolution was not only about teaching evolution but also about freeing education from the authority of the past, and its functionless curriculum. Huxley's support for the liberal educational agenda was foremost a rebellion against the authority of tradition, which still dominated English education.[114]

A curriculum which teaches nature's laws as interpreted by science has a dual purpose. First, it offers tools for individuals and society to progress materially, given that progress in Huxley's eyes is a function of "mastery over the forces of nature."[115] Second, by knowing nature's laws, it gives a clearer understanding of how to live. For example, in discussing a scientific education as no less of a liberal education than the Renaissance-based British humanist education, Huxley writes:

> We cannot know all the best thoughts and sayings of the Greeks unless we know what they thought about natural phenomena. We cannot fully apprehend their criticism of life unless we understand the extent to which that criticism was affected by scientific conceptions. We falsely pretend to be the inheritors of their culture, unless we are penetrated, as the best minds among them were, with an unhesitating faith that the free employment of reason, in accordance with scientific method, is the sole method of reaching truth.[116]

It is unclear if these two purposes in teaching nature's laws, the practical of material progress and the theoretical of knowledge and truth, are, in fact, distinguishable from one another. Huxley's discussion of technical education is indicative of the blurring in the distinction. On the one hand, he advocates that a technical education is no different than a good education—that is, a good liberal education which includes the natural sciences, the humanities, arts and language: "The education which precedes that of the workshop should be entirely devoted to the strengthening of the body, the elevation of the moral faculties, and the cultivation of the intelligence." Huxley adds, however, "and, especially, to the imbuing the mind with a broad and clear view of the laws of that natural world with the components of which the handicraftsman will have to deal. . . ."[117] Here, and elsewhere, his vision of education slips into an instrumental materialism. Huxley's curriculum, while not limited to the natural sciences, is at times driven by a vision of harnessing knowledge for the sake of material progress.[118]

While Huxley moves back and forth from a view of knowledge which is materially driven to one which is the basis for human self-understanding, it is clear that all knowledge needs to have a functional value. Greek, Latin, ancient history and literature should be maintained as part of the curriculum, but should be taught in order to teach lessons that are relevant, and not as obscure edifices, irrelevant to contemporary life.[119] Even the arts, as he understands them, have a function, as "the serene resting-place for worn human nature."[120] Huxley's curriculum is not limited to the sciences, but his commitment to a larger curriculum is always based on justifying its use-value to human beings.[121] Its utility is often defined by its importance for material progress; it is always evaluated for its functional relevance.[122]

Still, although Huxley, like Spencer, tries to reduce the liberal arts curriculum into the same functional criteria with which he sees science, he nevertheless recognizes that the liberal arts/humanities offer something different. Huxley never argues that the curriculum should swing from an all-classics to an all-science curriculum. He believes that they need to be balanced.[123] Although in his earlier period the monistic notion of a metaphysical worldview being grounded in our view of physical phenomena seemed to navigate his thinking, I think that his emerging dualism ultimately explains his educational philosophy more effectively. Science never covered everything. It explained nature's laws, which is the material world on which the cosmic process is based and on which material progress is dependent. The world of aims and of purpose, however, in the end remained outside of its reach. In describing the goals of education, Huxley states simply:

> No educational system can have a claim to permanence, unless it recognises the truth that education has two great ends to which everything else must be subordinated. The one of these is to increase knowledge; the other is to develop the love of right and the hatred of wrong. . . . [124]

Science increases knowledge, but values remain outside of the scientific realm, which is why the Bible remains inside his curriculum. Modern life has been strong on material knowledge, but cultural values are to be found elsewhere. Culture, through reason, transcends the material world; it can change one's character, freeing it from its baser, natural tendencies.[125] The human being is not a slave to his/her passions; s/he can conquer them. Natural differences between human beings can be neutralized. Women, attitudes about them always being a bellwether for how nature and society are seen to interact, can transcend their biology and essentially neutralize the inequality of childbearing.[126] As Huxley states clearly, "The duty of man is to see that not a grain is piled upon that load beyond what Nature imposes; that injustice is not added to inequality."[127] Nature here is far from benevolent; it brings the burden of inequality. Justice, however, is the province of society. Education must navigate the tension between its role as the agent of nature's laws and its role as nature's adversary and combatant.

KROPOTKIN'S EDUCATIONAL PHILOSOPHY

The primary difference between Spencer and Huxley on the one hand, and Kropotkin on the other, is that Kropotkin defines the human being as, naturally, a social being. By doing this, Kropotkin creates a prominent place for culture and society within his evolutionary worldview. Whereas Spencer maintains a radical individualism as a normative description for organizing society, and Huxley rejects nature in order to embrace culture, Kropotkin continues Darwin's intuition, and describes a picture of cooperative, human society emerging out of nature. In the same way, then, Kropotkin's ideas about education will be focused on the interplay between human nature and culture.

Kropotkin was not as prolific a writer as Spencer or Huxley, nor was educational philosophy a central theme of his work. Still, in his major essay on education, Kropotkin explains the fundamentals of his educational philosophy, clearly based on his theories of mutual aid and ethical development. The name of the essay, "Brain Work and Manual Work," already lays out the key component of his philosophy: that mind and body have been separated in culture as in education, and that education needs to reunite the two so as to allow what Kropotkin calls "a complete education" (*education integrale*), and a complete human being.[128]

For Kropotkin completeness, or wholeness, stands in opposition to contemporary society, which has fragmented the individual, focusing on only one aspect of what it means to be human—the intellectual, rational side. Education's goal is to teach the whole human being, which for Kropotkin means teaching the human being as part of the larger whole. He believed that such a view was in stark opposition to education as it was conceived.

Education was divided into "brain work" and "manual work"—that is, theoretical studies for the more "intelligent," and usually well-off students, and technical studies to learn a trade for the less "intelligent," and usually poor students. Such a dichotomy reified, and continues to reify, the gap between the haves and the have-nots. Rather than demanding that all students receive a theoretical education, however, Kropotkin called for praxis between what he calls "eye and hand" education, and "brain" education. In the division of the social classes, both sides suffer.[129]

In Kropotkin's view, societal progress is stifled by the divide. He argues that the nineteenth century, far from being a century of technological progress, was actually one of stagnation. In the first generation of the scientific revolution, there were men who combined theoretical knowledge of their minds with practical knowledge of their hands and eyes. Men of science were skilled in crafts, and they knew how to use their hands. They were exposed to the world and were stimulated by it, and they were complete. Reason was not decontextualized, but rather integrated into the natural, emotional, experiential side of human life.[130] Rather than contrasting reason and emotion like Huxley, Kropotkin sought to reunite them.

But Kropotkin's view of the complete human being is in the larger context of being part of an interacting community. Artisans, although not trained in theoretical knowledge, were exposed to scientists and exchanged ideas and information. Society, in Kropotkin's eyes, was neither isolated nor stratified, and as such, the explosion of scientific knowledge came as a result not of solitary geniuses, but of an interacting community where neither theoretical nor practical knowledge had become specialized and compartmentalized.[131] Sociability is a central part of human life, and is critical to societal progress. The ideal was not that of atomized, heroic individuals, but rather a community of learners.

All this had changed. Scientists, Kropotkin claimed, have turned the divorce of theory and practice into a value: "they have raised the contempt of manual labour to the height of a theory."[132] They were "devoid of inventive genius, owing to their education, too abstract, too scholastic, too bookish."[133] Conversely, artisans were no more. Workers became manual laborers in specialized tasks which breeded boredom and subdued creativity. The wholeness of the individual and society which was critical for progress was destroyed.

The goal of education, therefore, is to repair the damage, and to reunite human beings and their communities with being truly human, as dictated by our natures. Human nature is not an isolated self, independent of others. Being truly human only takes place within the social context of the community, and eventually in identifying one's life with the life of others. The reuniting of brain work and manual work is not only making the individual whole, but making society whole, by reuniting the individual with the society:

> Be it handicraft, science, or art, the chief aim of the school is not to make a specialist from a beginner, but to teach him the elements of knowledge and the good methods of work, and, above all, to give him that general inspiration which will induce him, later on, to put in whatever he does a sincere longing for truth, to like what is beautiful, both as to form and contents, to feel the necessity of being a useful unit amidst other human units, and thus to feel his heart at unison with the rest of humanity.[134]

Such an educational philosophy makes sense coming from Kropotkin and his theory of mutual aid. Since society flourishes when the individual acknowledges his/her own well-being with the flourishing of others, education needs to encourage and nourish mutual aid, by connecting the individual to society. "Brain education" assumed an atomized and disembodied view of the learner, teacher and subject matter. It needed to be reunited with "eye and hand education," which assumed a social and embodied view of the educational process. The artisan apprenticeship, in Kropotkin's worldview, is a paradigmatic education.

Curriculum Implications of Kropotkin's Educational Philosophy

Two central components make up Kropotkin's curriculum. The first is science, as it is for Spencer and for Huxley. Science reveals the truth of the world, and is the key to societal progress. The second component, however, is different from Spencer's curriculum, and reflects the ideological differences between the two. Although Spencer ultimately sees human society evolving a cooperative, altruistic human nature, the central unit which drives the evolutionary process is the individual. In Kropotkin's cooperative model, it is the interaction between individuals and the emerging community which is the central unit. From a biological perspective, that led to his viewing the *group* as the central unit of selection.[135] From an educational perspective, that led him to focus the curriculum not on the individual pursuing nature's laws as expressed in his instincts, as was Spencer's direction, but rather on the individual's interaction with others.

The centrality of science to the curriculum, not only as an instrumental tool for material progress but also as the critical tool for mapping the path to human moral progress, was common among those who saw a connection between "is" and "ought." Science mapped out the objective truth of society, which was critical to any attempt to then develop a normative morality. Kropotkin was an enthusiastic advocate of such a position. Science held the key to understanding the nature of the world, and therefore, the tools and direction for moral improvement.[136] Therefore, science for Kropotkin was not only about decoding nature's laws to be harnessed for human progress. Kropotkin saw science as playing an alternative role: challenging the centrality of the individual, and of the human species, and placing him/her in the larger context of the cosmos. Science teaches that humans are part of a larger story, and that the individual gains meaning only through its connection with the rest of creation:

> Modern science has thus achieved a double aim. On the one side it has given to man a very valuable lesson of modesty. It has taught him to consider himself as but an infinitesimally small particle of the universe. It has driven him out of his narrow, egotistical seclusion, and

has dissipated the self-conceit under which he considered himself the center of the universe and the object of the special attention of the Creator. It has taught him that without the whole the "ego" is nothing; that our "I" cannot even come to a self-definition without the "thou." But at the same time science has taught man how powerful mankind is in its progressive march, if it skillfully utilizes the unlimited energies of Nature.[137]

There is a religious aspect to Kropotkin's portrayal of science, although he would not define it as such. Science teaches the connection between living beings and the meaning which emerges as a result. In this sense, Kropotkin's two curricular components are complementary. Science does not simply teach the laws with which to harness the mysteries of the universe for society's material and moral progress. It also paradoxically teaches that the individual is connected to a larger whole, which is necessary for meaning in life. The atomized individual, separated from the rest of creation, is reunited with the cosmos through understanding the true message of science. Environmental studies today, for example, often accomplish such a pedagogic goal. The study of ecology allows students to become aware of their connection to the larger web of life.

Connection, then, is a central curricular motif for Kropotkin. Didactically, students cannot learn to connect simply by learning scientific facts. Science teaches the laws of nature, and that the metaphysical reality of our lives becomes significant only in relation to the larger whole. Humans need to experience such relations, however, and not simply cognitively accept that they are necessary. Therefore, Kropotkin's curriculum learns about life through life.[138] A doctor trained only through books can never develop the sympathies necessary to do his or her work correctly.[139] Only the learning of life through life—represented by manual work—will teach us to connect ourselves to others and, through that connection, to connect to our own humanity.[140] Those connections are not to be instrumental, but ones which nurture our innate sympathy. In order for sympathy to be expanded, the student needs to build such connections with others as an equal, and not as a superior coming to observe the working class.[141] Only by understanding the fundamental connections between individuals as part of the same human species, and transcending the boundaries of the ego which allows the student to imagine him/herself as different and superior to the worker, can such a bond be made.

The result of such a union is the fusing of sympathies for others with scientific knowledge, and the relinking of brain work with manual work.

Doctors studying only through books lose sight of the human being in their care, and reduce medicine to diseases which need to be combated through medical interventions. If their training would include work with patients in the places they live, they would understand what Kropotkin claims every good nurse who works with people knows, that disease is part of a larger social context—for example, poor hygiene, conditions of poverty, lack of education—and not an isolated, individual occurrence.[142] Sympathy for others, nurtured by full participation in the manual life of the world, contextualizes the book knowledge of science.

Pedagogical Implications

It follows that Kropotkin's pedagogy would focus on this interrelationship between "manual work" and "brain work." Learning takes place when both are present. Technical education has divorced manual work from its intellectual context; academic education has divorced brain work from its physical embodiment. Pedagogy needs to fuse the two. Kropotkin, however, clearly believes that the starting point of learning is on the practical, rather than the theoretical, side of the dialectic. In science, this means first the experiment, then the theoretical understanding, not the other way around.[143] The scientific method embodies such a pedagogy. Experiment and discovery emerge from confrontation with practical matters of the real world, and are only later distilled into theoretical laws.[144] Science, rather than a fixed methodology of applying theoretical laws, is an art in which practical matters raise questions to the curious who seek creative solutions.[145] Kropotkin's problem-solving pedagogy echoes Dewey's.

Children are naturally curious, and book learning stifles that curiosity.[146] Without interaction with the world, the curiosity which is the catalyst for learning disappears. The divorce of academic learning from manual work results in the extinction of curiosity, and consequently of discovery and scientific progress. Rote memorization is the bedfellow of book learning. Information becomes something memorized instead of discovered. Geometry, for example, is taught through the memorization of theorems, instead of their discovery through confrontation with real-life situations.[147]

The alternative, obviously, is a pedagogy based on interaction with the real world. Geometry needs to be taught not as assumptions and theorems to be memorized, but as a tool with implications for actions. It needs to be taught in places which make it relevant, for only then will it be truly understood and not simply memorized.[148] Technical education, far from being divorced from brain-work, is where theoretical ideas are to be discovered and applied. Experience and discovery are the central characteristics of Kropotkin's pedagogy of education, which stands in radical opposition to the ruling pedagogy of the schools as he saw it:

> By compelling our children to study real things from mere graphical representations, instead of *making* those things themselves, we compel them to waste the most precious time; we uselessly worry their minds; we accustom them to the worst methods of learning; we kill independent thought in the bud; and very seldom we succeed in conveying a real knowledge of what we are teaching. Superficiality, parrot like repetition, slavishness and inertial of mind are the results of our method of equation. We do not teach our children how to learn.[149]

CONCLUSION

Kropotkin's most obvious uniqueness lies in his description of the character of the natural world, what I have called earlier the first circle of analysis. He seeks to undermine the Social Darwinists' description of the natural world, which they claim to inherit from Darwin. Both Spencer and Huxley adhere to the Social Darwinist description of nature as selecting for individualistic, selfish and competitive behavior. Kropotkin utterly rejects this description of both Darwin's theory and of Darwinism in general, pointing out Darwin's belief that a social instinct is central to the social species and the development of morality. Kropotkin sees mutual aid and sympathy as the central factors underlying successful survival strategies in evolution.

Because Kropotkin, like Spencer, sees no significant barrier between the first circle of nature and the second circle of human nature, nor between the second circle which describes human nature and the third circle which prescribes human nature, the implications of such a cooperative, social description of nature for educational philosophy are significant. Like all the progressives, Kropotkin argues that education needs to return to nature's way. Spencer, Huxley and Kropotkin, for example, all believed that educational pedagogy should be experiential, rather than rote learning; one should learn about nature, in nature, naturally. Human beings are not passive, needing to be stimulated, but naturally active, curious beings. Return them to a natural setting, and learning will take place. Education should also pay attention to the natural ways that children develop, exposing the child to that learning which is "natural" for that stage of development.[150] One can see here the roots of a child-based curriculum. However, whereas Huxley eventually saw human nature's way being at odds with the rest of nature's way, and therefore reasserted the centrality of rationality as the focus of the educational enterprise, both Spencer and Kropotkin saw nature's way being a reconnection of reason with the emotions, which represent nature's law. Spencer often saw this as an education of both mind and body, maintaining their dualism and suggesting that education needs to redress the imbalance between them. Kropotkin, however, moves closer to breaking down the dichotomy altogether. "Brain work" cannot be truly brain work, that is, the proper development of reason, without "manual work," that is, without its expression through the body. Reason without body is abstract, detached, and decontextualized; body without reason is alienated labor and dehumanizing. While Spencer sees nature's laws as reconnecting us to our emotional, animal side, Kropotkin sees nature's laws as fusing these two aspects of what it means to be human, which are in fact one. Kropotkin foreshadows the rebellion against the rationalist assumptions of educational philosophy which are so prevalent in our day, while searching for a language with which to explain a more complex mind–body relationship, one in which rationality is embodied within, and can only be explained and cultivated as part of, our emotional lives.

The uniqueness of Kropotkin's position can be seen in the nature–culture divide, as well. Spencer's educational philosophy is based on a radical individualism, standing apart from the dictates of culture. Kropotkin, however, argues that human beings, by nature, are social. Culture is an extension of nature. For Spencer, natural learning meant primarily to learn from nature, although his Lamarckian views allowed nature's laws to develop in historical time. For Kropotkin, it meant no less learning with, from and for others. A return to nature's laws meant a return to a view of the human being as primarily a social being. Education takes place in social settings, and aims to reinforce our natural connections with one another. Our humanity is not expressed through developing our individual talents and abilities, but by building bonds outward into the world. The doctor becomes a doctor only when he ties the intellectual understanding of his discipline with the social/natural understanding that all work is about creating deeper, natural bonds of sympathy with others. These bonds are based on equality, recognizing that fulfillment in life is dependent on recognizing that life's fulfillment can only take place in the companionship of others. We are social, sympathetic beings, and our education needs to be rooted in celebrating and developing that most human trait. In this, Kropotkin importantly distances himself from the romantic assumption that the human being is, at its core, an individual separate from society.

Kropotkin can, like Spencer and Huxley, and the progressives in general, be viewed as building an educational philosophy that can all too easily slide into a material view of humanity, focusing on industrial education and material progress. The constant focus on relevance opens the door for all knowledge to be judged by its use-value. However, Kropotkin, like Spencer and Huxley, also holds a richer view of what it means to be human than a straight materialist view would seem to suggest. For all of them science is the key to progress, centrally material progress. However, for Spencer, science also unlocked the laws of nature which have been hidden from our instinctive nature by culture. Science taught nature's laws to which human nature must return. For Huxley, scientific method was the application of reason, human nature's most distinctive trait. Science taught reason. For Kropotkin, science taught humility. It taught that the human being's place was not as an isolated ego in the world, but as a being connected to the larger evolutionary story, whose life meaning can only be uncovered in the context of the whole. This is a theme which returns in Dewey, and with Midgley in second-generation Darwinism. Science, therefore, while unlocking the secrets of the material world, also symbolizes for Spencer, Huxley and Kropotkin a fuller view of humanity.

At his worst, Kropotkin, like Spencer, sees "nature" and "science" as eclipsing "culture" and "tradition," an argument leveled against the progressives, in general. Tradition and culture are suspect, as they are unexamined by the new tools of rationality and science. Cultural knowledge becomes unscientific knowledge, filled with prejudices and unexamined truths. Only Huxley, because he ultimately holds onto a binary view of nature and culture, values an education of morality rooted in a classical education. Having abandoned natural law as a normative system, the only one of the first-generation Darwinists to have done so, and simultaneously being a self-proclaimed agnostic, Huxley's dualism exists without normative guidance from nature. Huxley remains suspended between his Humean allegiances and a Victorian Christian religious ethic. He sees both the Bible and Renaissance humanism as curriculum anchors for teaching normative values. He is one of the first spokesmen for the strategy of separation of facts from values, where facts are material and accessed through science, and values are seen to be subjective and found in cultural norms.

All of them, then, are in a sense positivists. While Spencer and Kropotkin allow their view of nature and science to eclipse cultural knowledge,

Huxley simply relegates culture to the subjective, thus implicitly privileging science with its objective knowledge. All three see science as a mirror of the world as it really is. Yet, in Kropotkin's shift of the description of human society from a collection of atomized, individual beings bound together by a social contract, to an interdependent one of social cooperation, he nevertheless suggests a view of science which stands inside the social experience of human beings, without pretenses at a God's-eye view of reality. Science is a socially embodied activity, and therefore cannot be seen as independent and transcendent of human interpretation. Although Kropotkin never articulated such a hermeneutical view of science and its consequent implications for our understanding of human nature, it is not a stretch to recognize that Kropotkin's philosophy moves us away from an unproblematic positivist interpretation of the implications of evolutionary theory for human nature. It is left to the fourth exemplar of first-generation Darwinism to fully articulate such a view into a Darwinist educational philosophy. Rather than describing some essential, material, objective reality, John Dewey's philosophy sees human beings and their description of the natural world from within the human social context, what I believe is a necessary ingredient in a sophisticated application of Darwin to society and education, and for understanding the interactions between humans and the non-human world.

CHAPTER THREE

Dewey's Darwinism

Human Nature and the Interdependence of Life

Dewey, like many of his fellow pragmatists, was a Darwinist. He believed that Darwin's theory of evolution had radical implications for our understanding of human beings, and therefore for educational theory and practice. Surprisingly, then, very little attention has been paid to looking closely at the ways in which Dewey, who was arguably the most important American intellectual of the early twentieth century, understood Darwin. Contemporary boosters have reclaimed him for a variety of reasons—often because of his commitments to what is called today a communitarian ethic, and his well-argued critique of the atomized, competitive view of the individual as a threat to democratic society. What is neglected, however, is that his view of the individual and of community rests on his Darwinian view of human nature, which is the basis for his philosophical system.[1] Dewey clearly argued early on that Darwin forms the organizing principle of his philosophy, and nothing that he subsequently said suggests that he repudiated that belief.[2]

My claim, therefore, is that to understand Dewey one needs to understand his view of evolution, and through it, his extremely robust view of human nature. For Dewey, like Kropotkin, this leads to understanding that human beings are part of the natural world, not some disembodied mind, that the essential ingredient of their natures is sociability, and that sociability therefore structures our understanding of the good for humans. Although Dewey is often seen as negating essences, and thus essentialized views of human nature, I think it is more proper to think of Dewey as reconstructing a more flexible, but no less robust notion of what it means to be human, one based on Darwin's biology rather than Aristotle's. Midgley also embraces and expands on such an approach, as we shall see, but Dewey is still the foremost philosopher to have built an educational philosophy on a Darwinian-constructed *telos*.

CHANGE AND GROWTH ARE THE ESSENTIAL FEATURES OF DARWINISM

As is well known, Dewey understands the concepts of change and growth as the central characteristics of the natural world as described by Darwin. For Dewey, there are no essences in the world, but rather a reality that is constantly changing and evolving.[3] In Dewey's eyes, this is the true revolution that Darwin brought about, one which rejected an Aristotelian biology, and thus philosophy, of essences and constants.

Dewey describes the Aristotelian worldview, which dominated philosophy from Aristotle's time until Darwin's, as rooted in Aristotle's biological theories.[4] Just as a species is fixed and permanent, and has its own aim and purpose to grow into its completed and most perfect form (for example, the well-worn metaphor of the acorn growing into an oak tree), so too nature and knowledge are fixed and permanent. Change and flux are illusions which must be penetrated in order to see the ends which are absolute and forever true.[5] It was, in fact, this order and purposefulness of both nature and knowledge which, according to Dewey, allowed for science, since order allows predictability. Religious beliefs were then similarly constructed based on Aristotelian assumptions.[6]

Darwin, of course, through the success of his scientific theory, did away with Aristotelian biology. Dewey believed that the first and most central aspect of the Darwinian revolution in philosophy was the substitution of the notion of static and fixed purpose in the biological world with the notion of change.[7] Nature, far from static, was a changing reality in constant flux. For Dewey, since nature and philosophy, that is, physics and metaphysics, were linked, such a dramatic change in scientific theory must have significant implications for philosophy, and religion as well. Indeed, many concluded that such a loss of essences meant the loss of meaning, since all was random and in flux.[8] Dewey, however, did not see life as directionless and devoid of meaning. His naturalism, devoid perhaps of a purposeful end, was nonetheless a project aimed at asserting aims, purposes, and meaning into the natural world, and through it, to human life.[9]

Change for Dewey had two significances. The first was that meaning was not to be found outside of the present moment in some idealistic ends and purposes. Meaning is found in the interaction between human beings and their world. While ultimate questions about the meaning of life lose their significance because, simply put, we cannot know where evolution is going since it is contingent and therefore open-ended, in its place, rather than a world without direction and therefore devoid of meaning, is a world in which the particular and proximate context and direction of change becomes the object of meaning.[10] This is connected to the second significance for Dewey, which is that humans are not passive spectators in the natural world, but rather active participants in the direction of nature and history. In what appears to be a direct attack on Spencer's evolutionary interpretations, Dewey argues that there are those who profess evolution, but whose philosophical conclusions continue to advocate the static and closed universe of Aristotle and the Middle Ages, and with it the passive role of humans.[11] Dewey, however, views evolution as open-ended, subject to the influence of human acts and deeds. Values are not absolute, independent of the human context, out there waiting to be discovered. They instead emerge from the meeting of humans and the world in the present, particular situation.[12] For Dewey, the concept of change shifted human attention from the future to the present, and from the general and ideal to the particular and contextualized.

Change is contingent, however, and can lead in many directions, not necessarily desirable ones. But Dewey's well-known optimism comes in part from his belief that the direction of evolution nevertheless suggests progress. Like his contemporary process philosophers, Dewey claimed that life was a process whose destination is not determined, but whose direction is suggested by the evolutionary concept of "growth." Darwin himself shied away from associating evolution with a valuative term such as growth, which could suggest evolutionary progress.[13] Dewey's notion of growth is nevertheless rooted in Darwin's view of natural variation, where different strategies can give one individual an advantage over others in the struggle for existence and reproductive success.[14] These differing strategies can be considered growth, according to Dewey, when they contribute to evolutionary success. Because any particular variation can only contribute to evolutionary success relative to the changing environment to which one adapts and exploits, growth is also relative to the changing situation, and contributes to the individual's evolutionary success only in the particular context of the moment, once again supporting his pragmatist philosophical view. A natural variation can in one environment be considered a contribution to growth, while in another can be considered detrimental to it.[15]

Because change is an essential feature of the natural world, successful adaptation in Dewey's view is ultimately dependent on the flexibility of an individual to adapt to a changing reality. The more flexible an organism's behavioral response to changing situations, the more likely the organism is to survive and prosper in a changing environment.[16] Part of the uniqueness of the human species comes from the openness of human instincts, which are not fixed but rather shaped through interaction with their given situation.[17] As the educational philosopher Daniel Pekarsky points out, Dewey's view of growth demands that education foster a certain *immaturity* in the student, never completely closing him/her into mature behavioral patterns which shut off possible alternative behaviors when a different situation, with different behavioral needs, is confronted.[18]

Such a definition of growth, however, like Darwin's idea of evolution through natural variation and selection, does not in and of itself supply the normative value which Dewey seeks. Growth in this definition is a biological term which simply means being better adapted for physical and reproductive survival. A slightly more aggressive lion could be more adept at successfully exploiting the situation for his or her own survival. Applying such a concept of growth metaphorically to human society, a businessman might be more successful at exploiting a business situation by being cunning or cutthroat; he has learned to adapt and successfully exploit his environment. Has he, therefore, experienced growth, in Dewey's terminology? I shall return to this question shortly, only after first laying out Dewey's understanding of the natural and the social, and their implications for human life, which forms the basis for Dewey's concept of growth as a normative idea.

HUMAN BEINGS CAN ONLY BE UNDERSTOOD AS PART OF THE NATURAL WORLD

Applying evolutionary concepts such as change and growth to human life presupposes that human beings are contiguous with the rest of the natural world. The notion that human beings were part of the evolutionary story, and not something foreign to it, was one of the major messages of Darwinism. As opposed to rationalist, idealist philosophy on the one hand, and transcendental religious philosophy on the other, Dewey, like other Darwinians, saw that the meaning of human life can be found only in the context of being part of, and at home in, the world.

This was not, however, to be taken as meaning that human beings were to be reduced to "mere" animals, implying a crass materialistic perspective, as Dewey claims his critics either explicitly or implicitly accused. It did mean that humans are to be understood solely in their context as natural beings. There was no need to go outside of experience to explain human existence. Although Dewey never builds on the science of ethology to defend his positions on human nature, he does contend that observation of humans and other animals should methodologically allow us to understand the uniqueness of the human species, as well as its similarities with other species.[19] Dewey opposed what he called rationalist or transcendental religious positions, which maintained the dualism of mind and body, but he also opposed a crass materialism which denied human uniqueness. He sought a middle ground between what William James had called tender-minded and tough-minded views of reality—between a transcendental idealism and a mechanistic, atomistic materialism—two poles which continue to define the debate on human nature today, and between which Midgley seeks, like Dewey, to navigate a middle way.[20]

Dewey's middle road attempted to describe human nature without reducing it to a predetermined and often unflattering essentialism. Dewey understood the concept of an embodied individual continuous with evolutionary history to mean that humans, like other animals, were not blank slates at birth, subject to the shaping and molding of socialization, but rather were active beings with needs and desires which interact with the world. Dewey strongly attacked the description of animals as a passive atomistic mass with no desires, aims or purposes of their own, ready to be molded by their surroundings.[21] Building on the Darwinian psychology of James and Mead, he saw the human being as an active bundle of instincts and desires from birth.[22]

The claim that there is an innate character to human beings from birth is central to Dewey's philosophy, and is its critical starting point. For him, it meant that the individual was extremely active, overflowing with desires and interests.[23] Dewey rejected claims that a child's activity had its starting point in outside stimuli. While the child responded to stimuli, his/her innate, evolutionarily needs were what motivated the interest. The notion of an innate personality, which only then begins to interact with the world, is for Dewey a resource with which human beings can resist the dangers of socialization, since the human being, therefore, has innate motives from birth independent of society.[24] It is a point that Kropotkin made, and Midgley has expanded.

Rationality, therefore, does not create our motives, but rather orders and structures them into a functioning whole. This point is central to Dewey and, as we shall see, to Midgley as well. The innate motives of human life are what structures the human good; they are the material from which a human life is constructed:

The conclusion is not that the emotional, passionate phase of action can be or should be eliminated in behalf of a bloodless reason. More "passions," not fewer, is the answer. To check the influence of hate there must be sympathy, while to rationalize sympathy there are needed emotions of curiosity, caution, respect for the freedom of others—dispositions which evoke objects which balance those called up by sympathy, and prevent its degeneration into maudlin sentiment and meddling interference. Rationality, once more, is not a force to evoke against impulse and habit. It is the attainment of a working harmony among diverse desires.[25]

Rationality works with the innate personality and motives which human nature provides, shaping them in relation to one another, defining the whole human being in terms of the proper "harmony" among its constituent parts. These parts, furthermore, are not neutral desires, to be evaluated according to some transcendental criteria of what is worthwhile in a life, and what is not. They define, loosely, the good for the species. Dewey maintained that philosophers who had divorced morality from human nature, seeing morality as something foreign to it, inevitably built a morality which was disconnected from who humans really were, and was therefore unrealistic.[26] This was not just a comment that meant that a disembodied morality, that is, independently and rationally conceived or divinely inspired, needed to take into account human psychology in order to understand what was a reasonable morality for human beings, and which ideals were unrealistic, given a human being's psychological nature; rather, it was a philosophical, normative statement: Human morality is an outgrowth of human nature, and is contiguous with it. Like Darwin's commentary on the bee—that if human nature were different, its morality would be too—Dewey is arguing for a view of morality which emerges from within the evolutionary story. Humans are not at war with their natures, trying to suppress the less desirable elements. They need to cultivate a balanced sense of their multiple characteristics in order to live a richly human life. The good for the human species, like all species, emerges from within the evolutionary story, and is not independent or opposed to it.[27]

Although Dewey explicitly rejects teleological thinking as a consequence of the philosophical implications of Darwin's theory, there is certainly a teleological impulse in Dewey's view of motives. Rockefeller is correct in pointing out that needs, desires and satisfaction of natural motives play a critical role in Dewey's worldview, much as it did in Aristotelian philosophy.[28] They are not simply descriptive of human nature, but rather form its normative base. Emotions, therefore, as for Darwin, Spencer and Kropotkin, form the starting point for building a worthwhile human life. Reason and emotion do not stand opposed to one another. Rationality works with the motives that are present in human nature, structuring them together, creating "a working harmony among diverse desires."[29]

While desires and instincts are already present from birth, however, offering the structure from which a life is to be built, Dewey rejects that they are already shaped and formed. That is, while Dewey believes that there is a strong hereditary aspect to human nature, it is largely potential, owing to its prereflective character.[30] While believing that there was a connection between natural, spontaneous activity and life's ends and aims, Dewey did not believe that they were synonymous. Innate tendencies can limit, direct and initiate activity, but they are not a closed system that has only one "natural" way of developing. The possibilities, while not limitless, are open to development based on "the uses to which they are put" or directed.[31] Natural growth is open to many paths based on the interaction between the innate, instinctual beginnings, and its interaction with experience. Dewey is the first of the Darwinists to attempt a more nuanced understanding of the connections between the natural and the social. Dewey, therefore, although rejecting the blank-slate view which maintains that essentially there is no such thing as human nature, also rejects an essentialist position on human nature, which sees the human character largely already determined at birth. He argues with two schools of essentialism, both of which are still prevalent today, which present two conflicting views of the essential human nature, and with it conflicting views on educational philosophy.

The first view sees the human being as a natural being, whose growth is dependent on developing in harmony with nature. In this romantic view, any departure from "natural law" endangers the natural development of the child into a healthy being, and is thus anti-educational. Education is dependent on allowing the individual to develop according to what nature has prescribed. The open school movement has roots in such a tradition. The second view also sees human aims and goals as being attributed to an innate, essentialist human nature, but that nature is described differently. Here, humans are described in classic Social Darwinist terms: competitive, belligerent and egoistic.[32] While some uphold these characteristics as necessary for social progress during preliminary epochs of cultural history (Spencer), others curse the baseness and immorality of human nature, but accept it as an accurate description of the complex structure of the human being (Huxley). This description of human beings has reemerged in the sociobiology debate of the late twentieth century, and is heavily criticized by Mary Midgley, just as it was by Dewey.

This second essentialist view of competitive, human society as a direct product of human nature is often seen to be far more pernicious than the opposing view that sees human society as a product of societal choices, as the blank-slate model does. Societal choices can be changed and modified, but human nature is seen as permanent. War, capitalism, sexism and racism are all explained/justified in the essentialist models by the deep structures of who we are, and can be restrained or celebrated, but not ignored. Science reveals the source of societal structures in our innate human nature.[33]

By attacking the one-to-one relationship between instinct and action suggested, for example, in the classic Social Darwinist story, Dewey rejects these essentialist explanations of the social emerging seamlessly from the natural. Instinct is open-ended and can suggest multiple behaviors:

> ... it requires an arrogant ignorance to take the existing complex system of stocks and bonds, of wills and inheritance, a system supported at every point by manifold legal and political arrangements, and treat it as the sole legitimate and baptized child of an instinct of appropriation. Sometimes, even now, a man most accentuates the fact of ownership when he gives something away; use, consumption, is the normal end of possession. We can conceive a state of things in which the proprietary impulse would get full satisfaction by holding goods as mine in just the degree in which they were visibly administered for a benefit in which a corporate community shared.[34]

Conversely, behavior is made up of a multitude of instincts. The claim, for example, that belligerence, which might very well be among the many innate characteristics in the human being, explains the existence of war in society, does not do justice to the complexity of the relationship between instinct and activity. Belligerence on its own is part of one's natural makeup. It can contribute to an aggressive act, which might be ethical or unethical behavior in a given situation, but is only one of many innate instincts (and educated habits, which are discussed later) which contribute to a given behavior:

> Pugnacity, rivalry, vainglory, love of booty, fear, suspicion, anger, desire for freedom from the conventions and restrictions of peace,

love of power and hatred of oppression, opportunity for novel displays, love of home and soil, attachment to one's people and to the altar and the hearth, courage, loyalty, opportunity of make a name, money or a career, affection, piety to ancestors and ancestral gods—all of these things and many more make up the war-like force.[35]

In Dewey's philosophy, therefore, the fact that human beings were a product of the evolutionary story had significant implications. On the one hand, it meant a rejection of the mind–body dichotomy which continues in both rationalist and certain religious philosophies. It meant a belief in an innate, inherited nature with desires and motives. However, on the other hand, while Dewey held that there was an innate character to species, humans included, he rejected an essentialist view that associated specific character traits and behaviors to the prereflective human being at birth. Human beings are born neither "good," only to be ruined by artificial culture, nor "bad"—selfish, competitive and individualistic. Although born with innate tendencies, they can be formed in a variety of ways, based on their interactions with the environment. Dewey's notion of habit is the place where he tries to articulate a middle position between essentialized, biologized and deterministic views of human nature, on the one hand, and blank-slate, socially constructed ones, on the other.

THE NATURAL AND THE SOCIAL: DEWEY'S NOTION OF HABIT

No concept of Dewey's so encapsulates his view of the interplay between the natural and the social as his concept of habit. For Dewey, habit stood exactly at the interface between instinct and socialization.

Instinct, for Dewey, is innate in human beings.[36] These instincts, however, are open and undifferentiated, often in conflict with one another, and they suggest different, often opposing, behaviors. Additionally, the human being is born into human society, and hence these instincts are immediately being shaped and structured by culture which predates the individual. As Dewey states, " . . . to say that some preexistent association of human beings is prior to every particular human being who is born into the world is to mention a commonplace."[37] In Dewey's understanding of language acquisition, for example, there is an innate tendency to communicate. That innate tendency, however, cannot be described as speech, but rather as cries. Like other innate impulses, it is "chaotic, tumultuous, and confused."[38] It is the meeting of the natural with the social which takes place immediately at birth, a social which simultaneously predates the individual's own particular nature, and that allows language to develop. Language is a social phenomenon which could not exist without a natural, innate impulse to communicate, nor without a social history that structured it and allowed it to develop over time.[39]

Language, therefore, is dependent on the social world to come into being. And yet, it becomes "natural" in that it eventually manifests itself as if it were a completely reflexive, non-reflective activity. It becomes a habit, which is virtually indistinguishable from an instinct, in that there is no rational reflection needed to initiate its use, and it is embodied and indistinguishable from a reflex.[40] Habit operates like instinct in that it sits deep within the human being and seamlessly acts in the world. As Dewey states: "When we are honest with ourselves we acknowledge that a habit has this power because it is so intimately a part of ourselves. It has a hold upon us because we are the habit."[41]

How do habits develop? Dewey's concept of habit is completely connected with his concept of problem-solving. Habits are a social/cultural phenomenon that emerges out of the interaction of human beings with the world, and the attempt to resolve the problems which experience presents. In Dewey's seminal essay, "The Reflex Arc Concept," for example, he speaks of the child's innate but unformed desires being drawn to a candle.[42] The child puts his/her finger in the flame and is burned. What was an instinctual reflexive movement of hand to flame has now become a problem, because the flame is an object whose meaning is no longer clear. While it once meant attraction, it now also means pain. And the next time the child's attention focuses on the candle, it is probable that s/he will no longer instinctually move her/his hand to the candle.[43]

Rational thought has to now negotiate between conflicting meanings of the candle, and to offer a behavior. Experience will then tell the child if the new behavior is satisfactory, combining the two meanings of the candle into a behavior that satisfies both. If the problem is then solved, and continual experience confirms the conclusion, the behavior can eventually return to being unreflective, and a habit is formed, this time of "instinctually" moving away from the candle.

While this might suggest that each individual develops his/her own habits through rational reflection when conflicting problems of meaning arise, Dewey in fact saw community as the central agent of habits.[44] Learning a language is obviously not the child's struggle to communicate through his/her own invention of language, but rather the passing on of a cultural heritage, learned through the trial and error of generations of individuals, and stored in the cultural habits of a people.[45] Culture, in

Deweyan terms, is the wisdom of a people's problem solving, collectively gathered and passed down over the ages. It has become habit, but it suggests centuries of trial and error, for the moment satisfactorily resolved.

Notice the centrality of socialization to culture in Dewey's philosophy. This is far from those progressive philosophies of education which sought to strip the individual of the habits of society and who often saw Dewey as their inspiration.[46] Dewey felt this was both absurd and impossible. The habits of culture are the collective knowledge of a social history, containing within them the solutions which that culture developed to the problems which experience presented.[47] Ignoring this experience is to go back to the cultural drawing board. Human beings are often unaware of the history of cultural habits, as they do not by and large reflect on cultural habits as they are transmitted from one generation to the next. They are only issues in transmission if they trigger a problem for the individual. Otherwise, socialization proceeds seamlessly.[48] The individual does not begin at the starting line, as it were, but benefits from a rich cultural history.

The danger of habit can be seen when it confronts a changed environment. Habits can lead to prejudice when individuals, raised within cultural assumptions, are exposed to different cultures with different assumptions. One culture's good and bad, narrowly but adequately structured for a particular culture and transmitted over the generations, could lead to prejudice by rejecting alternative habits as immoral. Similarly, since environments change and habits are static, habits by definition will not continue to be appropriate behaviors for changing circumstances. Habits embody cultural judgments made in particular situations. When these situations inevitably change, habits become dangerously inappropriate.[49]

Habits also have the danger of being confused with instincts, which are innate. Habits are cultural constructs designed from the meeting between human desires and the larger environment in experience. Habits can change. Human culture manufactured them from the inchoate initial human instincts, and human culture can remold them. The remolding, however, is limited by the structure of our innate instincts. Society cannot progress in whatever direction it likes. It is not infinitely malleable.[50] For Dewey, the two schools, one of unchanging human nature and one of infinitely malleable human nature, have it exactly backwards. Conservatives have tended to hold a fixed view of human nature, since they believe human society to be a natural outgrowth of fixed human nature, which cannot change. Progressives, however, hold a malleable view of human nature, in that they see society as a product of human choices, which can also be changed and improved. They assume that there is nothing permanent about the current social structure. Dewey claims, however, that it is human nature which is malleable and open to change, and human social structure which is conservative and most resistant to alteration.[51]

The sources of cultural change for Dewey are the innate instincts of human nature. Instincts are the starting point of habits, through their unreflective and disorganized interaction with the environment, and they are also the starting point for changing habits. This can only be because they stand independent of society. They are socialized through experience and culture into habits, but they remain as raw material that is capable of bursting through the habits of society, challenging and rearranging conventions.[52] For example, the innate sympathy to which human beings are predisposed has been shaped into cultural habits which, although perhaps expressing prejudice in a new social environment, also often maintain and nurture those innate sympathies which can reject the prejudice and reshape the habit. Culture can thus be restructured, to allow a more inclusive habit to develop. As the environment changes, creating dissonance between the habits of old and the reality of new, as problem situations reappear, it is the ever-present innate instincts which can explore and react to the new terrain.

HUMAN BEINGS ARE BY NATURE SOCIAL ANIMALS, AND CAN ONLY BE UNDERSTOOD THROUGH THEIR SOCIABILITY

Now we can return to the question of how Dewey turns growth into a normative concept. Dewey believed that natural motives form the goods from which a human life is to be formed, to be structured into a working harmony. By what criteria, however, are these often conflicting goods to be structured into a whole? What determines why, for example, aggression or selfishness, perhaps natural motives, should be less central to the vision of a worthwhile human life than compassion or sympathy? If we try to use the concept of growth as Dewey meant it, as an arbitrator of the good, emerging from the evolutionary story as the single normative criteria for (human) life, then why would aggression and selfishness be less successful in facilitating human growth than the social motives? The answer, for Dewey, is ultimately because human beings are social animals, and therefore human growth is dependent upon social values. An innate sociability was not only describing human nature, but also prescribing how a human life should be lived. Our sociability is the defining characteristic of human life. As a naturalist philosopher, the good was not something to be found outside of experience. We value sociability because sociability, with its attendant characteristics of sympathy and concern for others, is central to what a human life is, and therefore, should be.

Dewey believed that the idea of growth is heuristic, navigating the move from the potential of human life with its prereflective innate motives, to a life which maximizes the human potential. Perhaps Dewey's clearest articulation of just how individual motives were to be evaluated and arranged into a prescriptive whole was in his article entitled "Evolution and Ethics" written in 1898 as a response to Thomas Henry Huxley's seminal essay of a similar name four years earlier, based on his Romanes Lectures. Dewey, as an admirer of Huxley, was particularly distressed by Huxley's seeming recanting of his Darwinian faith, by his reassertion of the dichotomy between the innate motives of human nature and the prescriptive values of human culture. As discussed in the previous chapter, Huxley uses the metaphor of a garden to illustrate the ongoing battle between the cosmic process and the ethical process. The human garden, that is, society and its morality, stand in opposition to the natural world which surrounds it and which is contained in innate human nature, with its weeds always lurking to conquer the garden. Like weeds, human innate motives are destructive to the whole, and human society has to create alternative values in their stead, and to rationally build an artificial nature to resist the noxious innate nature which evolution has provided.[53]

Dewey rejects this claim about human nature and attempts to rework Huxley's metaphor of the garden. Arguing for continuity with the natural world, he sees the garden as a part of nature, emerging out of the materials and evolutionary history of the natural world. Nevertheless, although from nature, the rules within the garden are different from those which have functioned until now outside the garden. Whereas fitness in the natural world might at one time in evolutionary history have meant the competitive struggle for existence between individuals for survival, at least for human beings the environment is now primarily a social one. The struggle for existence will be very different, owing to the different nature of the human being and his/her environment. In the social environment, cooperation among individuals is the successful evolutionary strategy, allowing the garden as a whole to flourish. Dewey argues that in this new social environment, behavior that was once considered fit is now unfit, whereas what was once unfit is now fit. Competition destroys the social fabric, necessary to the flourishing of the whole. In Dewey's words, " . . . we have reason to conclude that the 'fittest with respect to the whole of the conditions' is the best; that, indeed, the only standard we have of the best is the discovery of that which maintains these conditions in their integrity. The unfit is practically the anti-social."[54] In essence, Dewey is suggesting a group selection perspective, where the well-being of each individual builds social cohesion, and therefore evolutionary success. Further, the social is aided by human self-awareness and intelligence. Because of consciousness, human beings are now capable of actively cultivating the garden.

But consciousness doesn't mean that humans develop new rules by which to cultivate the good of the garden as a whole; human rationality discerns the rules of the garden itself, and becomes a partner in creation, not a substitute for natural laws. Rather than a war against nature, the gardener works with it.

> We may imagine a leader in an early social group, when the question had arisen of putting to death the feeble, the sickly and the aged, in order to give that group an advantage in the struggle for existence with other groups—we may imagine him, I say, speaking as follows: "No. In order that we may secure this advantage, let us preserve these classes . . . We shall foster habits of group loyalty, feelings of solidarity, which shall bind us together by such close ties that no social group which has not cultivated like feeling through caring for all its members, will be able to withstand us." In a word, such conduct would pay in the struggle for existence as well as be morally commendable.[55]

Dewey's argument with Huxley is that the metaphor of the garden in fact explains the connection, and not conflict, between society's and nature's essences and aims. Society does not stand in opposition to nature but is a continuation of it; the gardener is not doing an artificial act but a natural one. The weeds are preventing the garden from becoming a more complete environment. The dilemma is not between nature and society, but between the parts and the whole. The gardener must control one part of the garden for the good of the whole. In this way, competition leads to stagnation since it prevents the flourishing of the garden. Only when one takes into account the larger picture can the greater whole flourish. Culture, then, is not anti-natural. Culture works with the material which nature provides but shapes it according to the well-being of the whole. Since human beings are intelligent beings, the shaping process is a conscious one, as some parts of (human) nature are preferred to others, according to their contribution to the flourishing of the whole. The cultural process, therefore, neither accepts nature as it is, nor is it foreign to it, but rather, it shapes nature according to a principle which nature itself provides.

For Dewey, then, human growth is dependent on understanding that growth for humans takes place in the context of the larger social and natural whole. Because humans are social beings, and because the environment to which humans adapt is a social environment, human growth is ultimately in the context of our interdependence with others. Although other animals might have grown through competition, adapting to the environment which they lived according to the tools which evolution afforded them, humans adapt to their environment, largely a social environment, according to the tools which evolution had afforded them—sociability. Human growth is dependent on recognizing that our physical survival and flourishing in our social environment can only take place within the context of our social natures. Selfishness cannot ultimately lead to growth, since it is non-adaptive to the social world in which human life is meant to flourish. For Dewey, therefore, growth can succeed as a normative concept when understood in the context of the social nature of human beings; only that which contributes to the flourishing of the full human being, that is, the social being, can be defined as growth.[56]

However, growth for Dewey was not an individual endeavor, independent of the society in which the individual is found. The individual exists and comes into being through experience only in relation to the world. Growth, therefore, cannot be simply about integrating the individual's instincts and habits into a greater whole. The individual cannot become a coherent whole in isolation from his/her society, since his/her self exists only in relation to the society, and not apart from it. We are interdependent with the rest of the world, and grow as human beings only insofar as we recognize this interdependence and act to integrate it into our character:

> The kind of self which is formed through action which is faithful to relations with others will be a fuller and broader self than one which is cultivated in isolation from or in opposition to the purposes and needs of others. In contrast, the kind of self which results from generous breadth of interest may be said alone to constitute a development and fulfillment of self, while the other way of life stunts and starves selfhood by cutting it off from the connections necessary to its growth.[57]

True growth, therefore, cannot take place independently of society and the needs of others. Dewey describes a widening circle of the self, in which the self continues to grow through its identification with others. It is here that Dewey's religious mysticism emerges. Not an otherworldly religiosity, but rather a growing experience of self as it connects to a larger whole, outside of itself, reminiscent of Kropotkin's similar identification with a larger whole. Egotism stunts individual growth since it fosters people living independently, or even at odds, with others.[58] People must choose to recognize that their growth is tied to their identification with the whole—the social as well as the natural. The border between self and environment blurs as our individual acts gain meaning as part of the greater whole. Our lives are continuous with the rest of the world, not separate from it.[59] Evolution shows us to be embodied, and that life's meaning is in our connections with the rest of life. And our natural sympathy gives us a good starting point for identifying our own personal growth in association with a reality beyond ourselves.[60]

It has been argued that, in retrospect, Dewey's philosophy could be considered communitarian in orientation.[61] Although Dewey's individual is an autonomous being with free will, s/he is born into a cultural situation, with habits and customs which contain the moral wisdom of generations. The individual is not a blank slate, but rather born into a natural and social context that frame the questions which are asked, and the possible solutions which are available. Character emerges and develops as the individual confronts conflicting meanings which the interaction between the self and the environment offers, and seeks to reconcile them into a unified whole.[62] And, continuing the reasoning of some communitarians, the resolution of inner conflict between desires can only be resolved in the context of being connected to a society. Our lives are embedded in our relations with others, and a moral life can only be expressed within the arena of the social world in which meaning is to be found.[63] Those that criticized Dewey for ignoring the inner life of human beings are partially correct, as Dewey's inner life is expressed only in the social world, in actions, commitments and identification.[64] Although Dewey's notion of growth never sees an end to the process, but rather sees an ever-emerging end-in-view which then redefines the next end-in-view,[65] he comes closest to understanding fulfillment as the voluntary identification of the self with the greater good:

> The final word about the place of the self in the moral life is, then, that the very problem of morals is to form an original body of impulsive tendencies into a voluntary self in which desires and affections center in the values which are common; in which interest focuses in objects that contribute to the enrichment of the lives of all. If we identify the

interests of such a self with the virtues, then we shall say, with Spinoza, that happiness is not the reward of virtue, but is virtue itself.[66]

As Steven Rockefeller has shown in his biography of Dewey, religiosity remained at the core of Dewey's philosophical worldview long after he abandoned his affiliation with organized religion. He not only reconciled his naturalism with religious faith; Dewey's naturalist philosophy supported a religious faith as it suggested that true human growth can only be found in the identification of the individual with the greater reality of the universe. In Dewey's words, "Within the flickering inconsequential acts of separate selves dwells a sense of the whole which claims and dignifies them. In its presence we put off mortality and live in the universal. The life of the community in which we live and have our being is the fit symbol of this relationship."[67]

LIVING IN THE WORLD: DEMOCRACY AS A NATURAL VALUE

Dewey's commitment to democracy as a value is an extension of his view of human nature. For him, democracy did not simply allow for social stability so that each individual could pursue his/her own happiness, but was rather a core value which expressed a basic faith in human potential for growth in community. Democracy was the political structure necessary for facilitating growth, and based on the constant pursuit of human growth, democracy would lead to the building of what Dewey called "the great community."[68]

Dewey's view of democracy rejected the separation of the private moral life from the public realm. Although he was careful not to undermine the role of the individual to freely choose with others the common good, the attempt to build a community based on a commonly shared view of the good which was simultaneously good for all individuals was what democracy was about. Human nature develops in a social context, and life's meaning is found in the interaction between the individual and the society, not in the privacy of one's own life apart from society. A society which allows the individual to explore how his/her own good interacts with the good of others, in order to build a society where each individual can seek his/her own good within and through the common good, therefore, is the political goal of our social structures.[69]

The recognition of the individual good with the common good is not self-evident. The human being must choose a direction of empathy as opposed to selfishness. Still, the individual is given tools to assist him/

her in identifying the furthering of community well-being as the path to personal growth. His/her abilities of reason, combined with natural sympathies cultivated and habitualized, allow Dewey his democratic faith—his belief that democratic rules can achieve a democratic way of life, where citizens are involved in a constant search to build an ever-more inclusive and enabling society. Dewey recognizes that there are dangers in such a process, that the good of the individual can be compromised in the name of the greater good of society. Nevertheless, Dewey does hold an optimistic faith that, given a fair chance, the needs of the individual can become one with those of the community.[70]

Dewey's faith in the potential of human nature to discover and articulate his/her growth by identifying with other individuals within the democratic community was stifled by the inequalities within society. Dewey saw "negative liberty," using Isaiah Berlin's phrase, as a precondition for the pursuit of "positive liberty," which was the true task of democracy.[71] As long as power was distributed unequally in society, and political access to decision making was unequal, a true conversation could not develop, since the goal of the democratic conversation was building a society which equally fulfilled each individual's pursuit of the good life through and with the collective. Without an equal conversation, the society which develops will not serve the needs of all its citizens. Democracy must first allow each individual equal footing, equal access and equal opportunity. Without that, no real conversation can take place.

A society based on inequality will not only fail to fulfill the weaker in the society, but the stronger, while better off materially, will also suffer. They also cannot grow as individuals so long as they relate to others selfishly in a competitive relationship, where one individual is subservient to another, rather than in a cooperative relationship, in which each acts for the benefit of others and for society. Without equality, human beings cannot pursue their potential as human beings, since humans are dependent on one another for their growth.[72] The pursuit of growth is dependent on the ability to step out of his/her narrow perspective and be exposed to other perspectives which would force an individual to expand his/her own view to be more inclusive. As long as conversation in democracy is pursued with individuals or groups entering the conversation to control rather than to learn, growth cannot take place. This was not a true conversation, but rather another version of oppression, less explicit than oppression in totalitarian and authoritarian regimes, perhaps, but nonetheless abusive and repressive.[73]

Although Dewey identified such inequalities in manners of race, ethnicity and sex, it was inequality in the economic sphere which most troubled him. For Dewey, "industry and commerce have more influence upon the actual relation of human groups to one another than any other single factor."[74] In Dewey's view, industry's primary task was not to amass wealth, certainly not for the benefit of a particular class. This view of business and industry divorced economics from the moral sphere, as economics were treated as a means to amass capital. Dewey saw business as another experience which either leads to human growth or which atrophies it. In Dewey's view the role of economic life was its contribution to the individual and to society as a whole. Economics cannot be divorced from questions of meaning:

> From the moral standpoint, the test of an industry is whether it serves the community as a whole, satisfying its needs effectively and fairly, while also providing the means of livelihood and personal development to the individuals who carry it on.[75]

For Dewey, the economic system was accountable on two regards. First, it was based on the exploitation of one class by another, and therefore prevented an equal playing field, a necessary precondition for democratic conversation.[76] Second, it denied both the exploiters and the exploited personal growth by relating to economic life solely as an instrument for making money.[77] While the purpose of economic activity is to provide livelihood, it is also to foster the growth of the individual and society. Dewey was not a Marxist, but he was highly critical of laissez-faire capitalism, which he understood as the economic expression of that same selfish worldview that sees the meaning of life expressed and pursued in competition with others.[78] For Dewey, a democratic economy must work according to different assumptions about human nature. The economic life should be built on a moral worldview which sees human nature as fulfilling its purpose of growth through connections with others.[79] Rather than separate from morality, the economic life in a democracy is to be the most profound expression of democratic values. A democratic economy must avoid exploitation and encourage equal status. For Dewey, workers needed to be involved in decision-making in the workplace in order to avoid alienation between means of production and their ends.[80] A democratic economy must supply meaningful work for its employees. A workplace which stunted human growth denied the worker the ability to become more fully human. And finally, a democratic economy must produce goods that will contribute to the growth of society and the individuals which comprise it.[81] The economy of a society was the central arena for the interaction between individuals and their society; what is produced is a reflection of what is valued. Only when a level playing field is created, when there is mutual respect and equality for all citizens, and when economics reflect social values and not profit-making for the few, can the real work of democracy begin—the democratic conversation which aims at building a society together.

Democracy for Dewey was not a technical act at the ballot box of majority rules, or a question of protecting the rights of the individual and reducing suffering in order to pursue a life however one saw fit. It was, rather, the political and cultural manifestation of the pursuit of the good, and Dewey had a clear notion of where the good originated from, and how to pursue it. With an optimistic realism about human nature, Dewey believed that Darwin taught that change and growth were the underlying realities of the universe; that human beings were deeply embedded in this world, with instincts which were flexible enough to grow in many different ways but with an innate sympathy for others which was a good starting point; that meaning emerges in relation to the world, and that habits are our culture's translation of meaning into human behavior; that human growth would only take place in the context of our connection with others, creating an ever greater whole; that democracy was not only a system of government which would allow growth to take place, but that it was a value which embodied the building of a common good for all its citizens, in a cooperative and respectful conversation among autonomous and equal individuals.

While Dewey stated that such a worldview undermines an Aristotelian view of *telos*, I have instead tried to suggest that Dewey was recovering such a view. Abandoning ultimate ends, Dewey managed, through his view of human nature, to reassert what can be called proximate ends into human life, ends which emerge from the evolutionary story, and are robust enough to offer an Aristotelian view of the good for human beings and society, based on our innate natures. It is from these ideas of the meaning of life emerging through our interconnectedness with the world around us that Dewey buildt his educational philosophy.

DEWEY'S DARWINIAN EDUCATIONAL PHILOSOPHY

Like the other Darwinian philosophers of education, Dewey's philosophy of education was consistent with his view of human nature. Dewey's goal of education was clear: the continuing growth of human beings and society through the fostering of individuals and a society which pursues growth and has the tools to obtain it. Dewey's philosophy of education is indistinguishable from his general philosophy. For him all of life is about education, and although he recognized the role of education as an institution, and devoted himself to its development according to his philosophical theories, in the larger picture every experience, every problem and every moral choice confronts the individual with an educational opportunity.[82] How one responds to the challenges of living determines whether one grows or stagnates. Achieving growth in life demands taking the long view—making choices that would not only facilitate immediate growth, but, more importantly, open avenues for additional growth in the future. Growth is not only the immediate growth at any particular moment, but choices which open one up to further growth:

> That a man may grow in efficiency as a burglar, as a gangster, or as a corrupt politician, cannot be doubted. But from the standpoint of growth as education and education as growth the question is whether growth in this direction promotes or retards growth in general. Does this form of growth create conditions for further growth, or does it set up conditions that shut off the person who has grown in this particular direction from the occasions, stimuli, and opportunities for continuing growth in new directions? What is the effect of growth in a special direction upon the attitudes and habits which alone open up avenues for development in other lines? I shall leave you to answer these questions, saying simply that when and ONLY when development in a particular line conduces to continuing growth does it answer to the criterion of education as growing. For the conception is one that must find universal and not specialized limited application.[83]

Such growth could only take place in cooperative acts within and for society.

How society is structured, moreover, so as to allow growth to take place for all of its members is a necessary, although perhaps not sufficient, ingredient of education. The building of a democratic civic society is a critical goal of Dewey's educational philosophy no less than the curriculum in the school. Schools, for example, could teach that life is about sympathy and cooperative ties, but when society teaches us that what is to be valued is the selfish pursuit of individual material gains, then selfishness, individualism and materialism become the true curriculum.[84]

Dewey has been accused of being a romantic, and if that means that he was optimistic about the potential goodness of human nature, then he certainly was.[85] Dewey distanced himself, however, from the classic romantic educational position, as articulated by Rousseau, who Dewey claimed saw the human being as innately good, and the role of education as allowing the human being to develop spontaneously, according to his/her nature. Dewey cited Rousseau arguing that there are three sources for education—"nature, men, and things."[86] By nature, Rousseau means our innate, human natures; by men, our social context within which we put our natures to use; by things, the environment in which personal experience unfolds. For Rousseau, nature, being predetermined and therefore not in our control, needs to form the basis and the aim of education, to serve the spontaneous development of the individual.[87]

Dewey agrees with Rousseau in the limited sense of the critical importance of adapting educational methods to the innate, psychological make-up of human beings. Understanding the instinctual base of human nature is critical for education in that all educational practice must take into account its make-up, and work according to its contours. This is part of the progressive assumption of a child-focused curriculum. Although it can be developed in many different ways, instincts create the context in which education must operate: "the instinctive activities may be called, metaphorically, spontaneous, in the sense that the organs give a strong bias for a certain sort of operation,—a bias so strong that we cannot go contrary to it, though by trying to go contrary we may pervert, stunt, and corrupt them."[88] Furthermore, like Rousseau, Dewey believed that although there is an innate human nature, even that starting point varies. Each individual has a distinct temperament. One could not, in broad strokes, for example, talk about girls or boys, blacks or whites. Each individual is unique, and each has unlimited potential for growth. Uniformity of education did indeed suppress the uniqueness of each child by offering a one-size-fits-all pedagogy, or a stereotyped view of limited potential.[89]

However, as seen, Dewey rejected this view of a definitive innate human nature which determines the aims of human life. Although there was a pre-reflective instinctual self, in reality it only becomes a self in relation to the world. The instinctive inborn human nature creates the structure for growth to take place, but not its aims:

> The point may be summarized by saying that Rousseau was right, introducing a much needed reform into education, in holding that the structure and activities of the organs furnish the *conditions* of all teaching of the use of the organs; but profoundly wrong in intimating that they supply not only the conditions but also the *ends* of their development. As matter of fact, the native activities *develop*, in contrast with random and capricious exercise, through the uses to which they are put.[90]

Dewey returns to his example of language acquisition to explain the difference. If one were to follow Rousseau's line of reasoning, language should develop naturally without cultural/social involvement, but that would only produce babbling. Innate nature, however, furnishes the qualities which allow language to develop, and it is then organized through the experience of culture.[91]

It is curious that Dewey's philosophy was interpreted in a Rousseauian spirit within the open education movement.[92] Dewey was attacked by critics for his rejection of tradition and culture, and with them the moral wisdom they contain.[93] Critics saw Dewey replacing the wisdom of cultural traditions with a child-centered curriculum with no sacred cows, everything open to inquiry and discovery, and attention focused on the future, rather than the past.[94] He can certainly be seen as being partially responsible for this interpretation, perhaps because in his attempt to offer a middle ground, he sees the danger of an unreflective and authoritarian tradition as far more dangerous to the immediate political educational reality than one which focuses exclusively on the child and his spontaneous development. Dewey, however, became cognizant of the dangers of a misinterpretation of his work, and in *Experience and Education* tries to refocus the progressive educational movement onto a middle position.[95] In spite of the criticisms, Dewey has a central place for tradition in his philosophy, and by extension, his educational philosophy. Habits contain the wisdom of the ages as they have shaped human instincts and given them form and content, and are critical to thinking, because thought is predicated on experience. As Dewey stated, only one who knows how to stand erect can know what it means and feels like to stand straight.[96] Our critical thinking is dependent on habits which have been formed through the collective experience of culture, and a central part of education is cultivating habits.

The other side of Dewey's educational philosophy, and where critics and supporters have often focused their attention, is when a new situation develops, and habits are no longer adaptive to the new environment. The uneducated individual is the one who is incapable of addressing the problem which this new reality presents, or who chooses a path which leads to the stagnation of growth. Dewey, like other progressives, believed that traditional education valued obedience to authority instead of reflective, critical thinking. Society, therefore, was destined to stagnate without opportunity for further growth.[97] The fact that Dewey placed his educational emphasis on problem solving rather than on the initiation of habits opened him to an interpretation which ignored initiation into a cultural tradition. It is probably more accurate to say that Dewey saw initiation into habits occurring without the need for formal educational involvement. Education's role was to reframe habits which needed reframing, while social situations would take care of the cultural initiation.[98]

The initiation into habits, their selective reframing through reflective problem solving, and their eventual return as improved habits was a constant process which involved relating a specific problem to its larger background. Dewey's philosophy of education was built on a dynamic of the integration, or what was earlier referred to as interpenetration, of one experience with another, and of one habit with another. He sought to educate for the integration of competing urges, or competing definitions of meaning, with a new, more inclusive definition that resolved the competition between conflicting tendencies. Dewey primarily aimed at educating the individual's character to be a more integrated self, both inwardly and with society.[99]

Dewey's curriculum was aimed primarily at fostering habits that were most conducive to further growth through a pedagogy of problem solving. Since he believed that the school needed to educate for citizenship in the world, problems needed to be real ones posed by life, and not artificial ones. Like Kropotkin, Dewey argued that learning needed to be based on experience, not on theories separated from the texture of life; school was not to be an institution separated from the world:

> I am told that there is a swimming school in a certain city where youth are taught to swim without going into the water, being repeatedly drilled in the various movements which are necessary for swimming. When one of the young men so trained was asked what he did when he got into the water, he laconically replied, "Sunk."
>
> . . . The only way to prepare for social life is to engage in social life. To form habits of social usefulness and serviceableness apart from any direct social need and motive, apart from any existing social situation, is, to the letter, teaching the child to swim by going through motions outside of the water.[100]

Students needed to be prepared for the world, not as a technical workforce, as some critics claimed, but as a contributing member in the conversation of building "the great community."

Because of his emphasis on growth and preparation for life, critics have also held that Dewey placed his emphasis on the future, rather than on the present.[101] The joy of childhood was consistently undermined for future goals; the future usurped the present. Dewey, however, was very explicit that this was not the case. "If I were asked to name the most needed of all reforms in the spirit of education, I should say: 'Cease conceiving of education as mere preparation for later life, and make of it the full meaning of the present life.'"[102]

Even as Dewey was aware of the functional need to continue to grow, that all ends are merely means to other ends, what he called an ends-in-view, and never an ultimate ends, still, the here and now could be experienced at any given time as having ultimate meaning.[103] Problem solving forced the individual to leave the comfort and seamlessness of the present, and to deal with the problem of life. Each moment, however, contains within it the realization of the moment's potential meaning, while simultaneously presenting a problem which demands reflection and choice. While education needs to foster growth, it also needs to help celebrate the meaning of the moment. Schooling, therefore, should not only be directed to the further development of the child, but should also allow the child to be who s/he is. Human nature is not only about becoming, but also about who the child already is. Dewey's extremely difficult pedagogic task was to allow the child to at once delight in his/her own being, in the nonreflective joy of the moment, and simultaneously to nudge the child to see within the moment the potential for further growth:

> The problem was to find "the forms of activity (a) which are most congenial, best adapted, to the immature stage of development; (b) which have the most ulterior promise as preparation for the social responsibilities of adult life; and (c) which, *at the same time*, have the maximum of influence in forming habits of acute observation and of consecutive influence."[104]

In the dichotomy of whether school should be play or work, present or future oriented, Dewey would argue that education is succeeding when the two meld into one, without being reduced into either.

Finally, Dewey's philosophy of education, true to his philosophy of the relationship between the natural and the social, tried to assert both the identity of the individual and, at the same time, the expression of his/her identity through ever greater identification with the social world and even the universe. He has been criticized for focusing on the social at the expense of the individual, and certainly he focuses on the public individual rather than on the private one, since the private individual does not exist for Dewey apart from the public. This is Ryan's greatest complaint—that Dewey's philosophy does not afford an understanding of the inner life of human beings.[105]

In the public realm, however, Dewey recognizes the difficulty of melting "individual" identity of cultures into a larger national identity. He was a nationalist, and did not believe that society should simply celebrate its newly found multiculturalism.[106] That would not afford a mechanism which would allow growth to bridge the gap between various identities, and allow the possibility of building an American identity which simultaneously allowed the expression of particular cultural identities. Similar to his model of the relationship between the individual and society, in which neither is reduced to the other and yet combine to become something different, immigrant cultures needed to choose to be American, and America needed to fully integrate a host of new voices into what it means to be an American.[107] A level playing field meant that all voices were to be participants in the building of "the great community," a community which doesn't suppress the autonomous individual, and submit him/her to the good of the larger community, but rather, through equal conversation, allows a community to merge out of the rich and variegated voices which make up America.[108] The goal of schools was to foster habits which would teach children how to be participants; the goal of education was to engage us in the unending process of building "the great community."

CHAPTER FOUR

Mary Midgley and the Ecological *Telos*

Darwinism in general, and Darwinist educational philosophy in particular, waned during the first half of the twentieth century. Its disappearance is partially connected to the waning of Darwin's scientific theory, which suffered from a lack of a clear theory as to the mechanics of natural selection which would confirm the larger hypothesis of how evolution took place.[1] Although genetics emerged as a theory in the early twentieth century with the rediscovery of Mendel and his theory, not until the 1930s and 1940s was it seen as confirming Darwin's theory of evolution.[2] However, in spite of its struggles, the scientific theory was not abandoned. In contrast, Darwinism—the application of Darwin's theory to the social sciences—was.

Carl Degler's *In Search of Human Nature: The Decline and Revival of Darwinism in American Social Thought*, documents the history and causes of the disappearance of Darwinism in America.[3] Degler sees the disappearance of Darwinism as a reaction to the Social Darwinist agenda, especially prevalent in America, of reactionary, often racist, social policies, which justified the growing economic and social gaps as the proper expression of natural, innate differences between people.[4] Franz Boas, the anthropologist, in reaction to the cultural assumptions that had become prevalent, worked to decouple behavior from nature, intelligence from race, gender gaps from sex differences, by showing the flexibility of human nature, and the biases in the scientific evidence, which "proved" innate inequalities.[5] Whereas Social Darwinism viewed human behavior and social structures as primarily emerging from differences in human nature, Boas, and increasingly the next generation of sociologists and anthropologists, saw human behavior and social structures as a product of human choices.[6] The pendulum swung from nature to nurture. The rise of fascism, and particularly Nazi ideology, using themes from Darwinism such as "survival of the fittest," eugenics and race theory, guaranteed the delegitimization of Darwinism.[7] A reactionary social Darwinism became associated with all attempts to apply Darwinism to social issues, in spite of Darwinism's progressive historical legacy. Cultural construction of human nature emerged as the predominant progressive position, while essential nature was seen as reactionary and racist.[8]

So when and how did Darwinism reemerge? Degler argues that it was scientific discoveries, combined with a growing sense among social scientists that the plasticity model was ignoring real insights from the biological sciences, often due to ideological assumptions of seeing plasticity as progressive, and innate nature as reactionary and regressive, which led to the shift back.[9] Although there were various benchmarks in the return of Darwinism to the social stage, for the larger public there is no doubt that the new Darwinism was born with the publication in 1975 of *Sociobiology* by the Harvard biologist E. O. Wilson, and subsequently *The Selfish Gene* in 1976 by the Oxford ethologist Richard Dawkins.[10] This is when Darwin's theory began to be reapplied in the public debate to social implications, and began what has been dubbed "the Darwin Wars," or what I call second-generation Darwinism.[11]

Mary Midgley, a contemporary British philosopher and important public intellectual, is a unique voice in the often strident debate in second-generation Darwinism, who began working on the issue of human nature as a central concept for philosophy already in the early 1970s. Since publishing her first book, *Beast and Man,* in 1978, Midgley has written scores of essays and books, exploring the interface between humans and nature, and its implications for human meaning and responsibility. Midgley's project began with her critique of the social science's unrealistic attitude to human nature. After reading the growing ethological literature, particularly Tinbergen and Lorenz, she became convinced that human nature, emerging out of its evolutionary history, was not merely a "dough-like product at birth."[12] Particularly the political left, in her estimation, had forged throughout the twentieth century a belief that social change was predicated on the ontological flexibility of human nature. Societal inequalities, the left contended, were the result of human choices. Any belief that there was an innate human nature suggested that social inequality was an extension of natural differences between rich and poor, black and white, and men and women, and was therefore justifiable.[13] Midgley saw this tying of a progressive political agenda to an unrealistic view of human nature as mistaken factually, philosophically and politically, as is discussed later.

According to Midgley, there were two different historical traditions converging in constructing what I have called the blank-slate position. The first, emerging from the empirical tradition and exemplified by behaviorism in psychology, denied the relevance of motives in the study of animal behavior and, in its most extreme form, denied that such motives even exist. Human beings could be conditioned and shaped in whatever direction society chose. They truly were blank slates, appropriate for social engineering.[14] For the behaviorists, this view of the human being was strongly linked to their notion of freedom, unshackled by innate nature. But, as Midgley argued, freedom is dependent on a unique human nature which is worth protecting and fostering. The idea that human nature is completely malleable undermines the notion of there being basic human needs, which serve as a basis for a vision of what human beings are, and therefore what society should be in order to serve and protect them:

> The notion that we "have a nature," far from threatening the concept of freedom, is absolutely essential to it. If we were genuinely plastic and indeterminate at birth, there could be no reason why society should not stamp us into any shape that might suit it. The reason people view suggestions about inborn tendencies with such indiscriminate horror seems to be that they think exclusively of one particular way in which the idea of such tendencies has been misused, namely, that where conservative theorists invoke them uncritically to resist reform. But liberal theorists who combat such resistance need them just as much, and indeed usually more. The early architects of our current notion of freedom made human nature their cornerstone. Rousseau's trumpet call, "Man is born free, but everywhere is in chains," makes sense only as a description of our innate constitution as something positive, already determined, and conflicting with what society does to us. Kant and Mill took similar positions. And Marx, though he officially dropped the notion of human nature and often attacked the term, relied on the idea as much as anybody else for his crucial notion of dehumanization.[15]

This point is critical to Midgley's argument, and is expanded later.

The second tradition which contributed to the plasticity position was the rationalist tradition, with origins in the Cartesian mind–body dichotomy and exemplified, in Midgley's view, by existentialist philosophy. Mind was considered to be where consciousness and thought resided, where rationality and morality lie, and was thought to be radically distinct from the body, instincts and the emotions. Whereas behaviorism "solved" the problem of human consciousness, and therefore of thought and motivation, by denying its relevance or even existence, the rationalist tradition continued to relate to consciousness as "the ghost in the machine," which is distinct from, and independent of, the rest of the natural world.[16] Our humanity, linked to our minds and consciousness, as opposed to our bodies and emotions, is tied to our rational thought, exercised through objective reason and free will. The rationalists assumed, according to Midgley, that any linking of the mind to the body, or of reason to emotions, would limit human freedom by tying it to the determinist, mechanistic processes of the rest of the natural world, which the body and its emotions were seen to represent.[17]

For Midgley, existentialism carries the argument to its absurd but logical conclusion. Having disengaged the human being from the natural world, and seeing free will as something which has been detached from any a priori claims on its decision making, it has simultaneously disengaged it from the context which gives free will and human life its purpose. Quoting the philosopher Nicholas Dent, Midgley reasons:

> One cannot represent all specific circumstances which impinge on an individual, all attributes ascribed to an individual . . . as limitations on, constrains upon, the self. For what then remains, to comprise the concrete actuality of the self's existence, is nothing, or almost nothing, a will without grounds, a power of choice without objectives.[18]

Existentialism has done that, by disembodying the mind and therefore ignoring the context and pathways within which human reason and free will operate.[19] Midgley claims that such a position is obscurationist, in that it obscures and obfuscates our ability to understand where consciousness comes from. It sees the human being as having miraculously appeared, independent of the evolutionary story. In her view, Steven Jay Gould, in spite of his Darwinism, is partner to this worldview, by refusing to acknowledge that being a Darwinist means that human beings have significant innate tendencies and motives prescribed by their evolutionary origins.[20] Gould and others, viewing what they call fundamentalist Darwinism as a scientific guise for reactionary agendas, continue the Boas tradition of decoupling culture from biology, only they argue from their authority as natural scientists.[21]

Midgley, contrary to the behaviorist and rationalist philosophies which she views as being anti-Darwinian, aims at presenting a Darwinian moral philosophy. First and foremost, she sees us as embodied beings, products of evolution, and understandable only in this context. Both behaviorism and existentialism in their own ways deny the lessons of Darwin, by denying that we can be understood like the rest of nature, through our evolved beings in their evolutionary context. Our motives all have a natural history, which make us not blank slates, but beings with personality from birth. Our rationality also has a natural history, and therefore cannot be seen as something distinct from instinct and emotions. Reason and emotions are all part of the same machinery.

And so Midgley set as her goal reconnecting philosophy with human nature, through a critical use of Darwinism as her guide. Just as she began her task, however, she found that although others had begun similar work, both their assumptions and conclusions disturbed her. If the one side of the pendulum had detached human beings from the natural world, the other side, in her eyes, had practiced a crass reductionism that described all of nature, human beings included, as part of a competitive, selfish and atomized existence, continuing the social Darwinist tradition. Dewey had also attacked these trends in the first generation. For Midgley, there are two paths in which the reductionism expressed itself, just as there were two paths which had rejected Darwin's insight of humans as embodied beings.

The first view, which Midgley saw articulated both by E. O. Wilson and especially by Dawkins, is the continuation of that part of the "Social Darwinist" tradition that saw life as essentially a competitive dog-eat-dog existence. Midgley wrote a scathing critique of Dawkins's *The Selfish Gene*, claiming that his description of the gene as selfish was a metaphor that Dawkins meant to extend prescriptively to human life. To Midgley, Dawkins was advocating a "biological Thatcherism":

> . . . romantic and egoistic, celebrating evolution as a ceaseless crescendo of competition between essentially "selfish" individual organisms, each making "investments" for its own separate advantage, organisms whose attempts to "manipulate" one another provided the whole dynamic of development.[22]

Midgley accuses Wilson and especially Dawkins of a sloppy and often confusing reductionism, borrowed from classical physics, in claiming that the psychologies and behaviors of living organisms can be explained through the idea of natural selection on the genetic level.[23] Dawkins argues that genes "behave selfishly," meaning that they are engineered through the selection process for self-promotion, and that this selection leads to the evolution of virtually all behaviors in living organisms. "Selfish genes" refers to the evolutionary process of maximizing one's own gene representation in future generations. As Dawkins himself points out, this obviously does not describe a process of choice among genes, which are not conscious entities with motives. Nor does it describe any psychological reality of the organism. One could argue that the "selfish" process of natural selection could produce altruistic, rather than selfish, motives and behaviors among a species.[24] It would simply be necessary to show that it would be evolutionarily advantageous for organisms to have altruistic behaviors, if one assumes that natural selection is responsible for the development of all behaviors. An altruistic psychology could have emerged from a "selfish" evolutionary process.[25] When E. O. Wilson called altruism the central problem of sociobiology, he was contemplating how natural selection, essentially a "selfish" process, could produce altruistic behavior. The "selfish" process of natural selection can logically produce a pluralism of psychologies and behaviors, each of which promotes genetic reproduction.

And yet, in spite of the fact that this point is well known to sociobiologists, they continually refer to selfish motives and behaviors among organisms, including human beings. Although the adaptationist agenda, where every motive has its evolutionary origin and explanation, should tell us nothing about an organism's motives and behavior as long as it promotes genetic reproduction, sociobiologists continually assume that it does. In their view, altruism is not an authentic, psychologically defendable position, but rather a self-deception hiding the true, selfish and competitive motivation for all human action. As the sociobiologist M. T. Ghiselin famously argued: " . . . given a full chance to act in his own interest, nothing but expediency will restrain him from brutalising, from maiming, from murdering—his brother, his mate, his parent or his child. Scratch an 'altruist': and watch a hypocrite bleed." [26] The biological description of the mechanism of natural selection as "selfish" therefore becomes the psychological description of the nature of life—selfish, competitive, aggressive and self-promoting. It is a confusion of categories.

Midgley, like Gould, does not accept the idea that the mechanism of natural selection can completely explain evolutionary processes. Even after clearing up the confusion between the origins of human nature and its character, it is important to Midgley that the evolutionary origins of human nature not be abandoned to the reductionist, determinist position. Natural selection does explain much of the evolution of life's character, although Midgley acknowledges that there are many epiphenomena which are not shaped by natural selection, but rather come along for the ride. Yet, in Midgley's eyes, like Darwin and as opposed to Gould, natural selection plays the predominant role in shaping our innate characters, and the innate character of the natural world.[27]

The second path of the reductionist position, also based in the Social Darwinist tradition, accepts the cultural assumptions of the "selfish gene" argument, but views them as cause for optimism. In the first path, the announcement that human beings are, at their foundation, involved in a "war of all against all," based on traditions ranging back to Thomas Hobbes at least, is a pessimistic, uncomplimentary view of human nature and life. Dawkins is clearly representative of this attitude, as was Huxley. In spite of the glibness with which our nature is advertised as being base at its core, Dawkins does not claim that this is good news. The second path of the "Social Darwinist" tradition, however, holds that this competitive selfishness, rather than being seen as a liability, should be celebrated as the path to progress. Rather than seeing the evolution of life as a contingent historical event, Social Darwinism sees the invisible hand of competition leading civilization constantly upwards, as Spencer did—what Midgley calls "the escalator fallacy." According to this view, "evolution is a steady, linear upward movement, a single inexorable process of improvement.[28]

Notice that Midgley's identification of the two poles which resist a proper Darwinism—the blank-slate position on the one hand, and reductionist social Darwinism on the other—are the exact same poles which Dewey identified as problematic in the first generation. Midgley is trying to navigate a middle path, one which advocates a robust view of human nature, and simultaneously rejects the reductionist and politically reactionary conclusions that are often drawn from such views. At its heart is a view of needs as central to the good of the species, that is, that our biological nature structures for us the basic outline of what can be defined as the good. Continuing the task that Dewey began, Midgley works to outline an Aristotelian view of the good from Darwin's theory of evolution.

INNATE NEEDS

Applying Aristotelian categories, Midgley believes that Dawkins and Wilson confuse what Symons calls ultimate and proximate causes of behavior.[29] They assumed that if natural selection is a competitive, individualistic process on the genetic level, then human motives emerging from such a selection process will be competitive and individualistic as well.[30] Although accepting a plurality of causes for evolutionary change, Midgley nevertheless accepts natural selection as the primary cause in structuring "the ultimate cause" of behavior.[31] Midgley's focus, however, is on proximate causes of behavior—the psychological makeup of human beings—which are perhaps a result of natural selection, but not reducible to its mechanisms:

> Motives have their importance in evolution and their own evolutionary history—but they have also each their own internal point, and it is virtually never a wish to bring about some evolutionary event, such as the maximization of one's own progeny. Confusion between the aims of individuals and the "aims" of evolution—if there can be said to be such things—is ruinous.[32]

For example:

> ... play no doubt has a function. It has been developed among human beings, as it has among the young of other intelligent species and sometimes among adults too, for evolutionary reasons which presumably have something to do with the satisfactory working of the higher faculties, with the need for practice in developing them, and with the sort of social interactions needed in a society which is much freer and less mechanical than those of the insects. It seems sensible to say that this tendency evolved because it has some value in promoting survival—that is, *as a result* of its having that value. But to say 'then it is only a means to survival' would miss the point entirely. *What* evolved was not only a tendency to act in certain ways, but a capacity for delighting in certain things, and thereby of taking them as ends. The ends of art and sport are now our ends.[33]

Because of this, far more than because of evolutionary theory, Midgley turns to ethological theory and observation to gain an understanding of human nature. Evolutionary theory describes the ultimate causes of behavior—what mechanisms shaped behavior—but ethology describes the proximate causes of behavior, the particular types of behavior which evolved in order to fulfill ultimate ends. Midgley is not primarily interested in the purposes of evolution, but rather in the purpose of human life. Proximate causes, and not ultimate causes, are where we find human needs and purpose.

How do we go about knowing human nature? While anthropologists, inspired by Boas, have focused on the differences between human cultures, Midgley argues that the differences rest on the backdrop of a commonality across all cultures which is too often ignored.[34] Characteristics such as smiling, nurturing the young and play, when spread throughout the wide spectrum of human cultures, point to universal characteristics rooted in a shared human nature. Similarly, characteristics shared with our nearest evolutionary relatives are signs of evolutionary origins for those characteristics. From these methodologies emerges the picture of a robust innate human nature, not something separate from culture, but forming the substrate on which culture is formed. As Dewey described, the human being, like all life forms, is not a passive creature waiting to be shaped by its environment, but rather from the moment of birth is programmed through evolutionary natural selection to actively reach out into the world. Echoing Dewey, Midgley argues:

> Anyone who doubts this should attend to the phenomenon of play. Neither human nor animal babies begin their lives at all as the mechanistic reactive theories would wish, by waiting passively to be acted on, and then imitating mechanically what is done around them. Instead, both equally take the initiative.[35]
>
> All this behavior is *active*; it is not in the least like getting caught in a snare; it will not break down into passive stimulus-response patterns. It is not mechanical but purposive, and the purpose is linked to lasting character traits expressing priorities.[36]

These motives, however, are not predetermined, independent of environmental influence. Ethologists have shown that the more social animals, as well as the more intelligent ones (and, as Darwin, Kropotkin and Dewey argued, there is a link between the two), have both closed and open instincts. The more intelligent the animals, the more likely they are to be dominated by open instincts.[37] This is far different from not having any instincts at all:

> When people such as Watson say that man has no instincts, they always mean closed instincts. They point to his failure to make standard webs or do standard honey dances, and ignore his persistent patterns of motivation. Why do people form families? Why do they take care of their homes and quarrel over boundaries? Why do they own property? Why do they talk so much, and dance, and sing? Why do children play, and for that matter adults too? Why is nobody living in the Republic of Plato?[38]

Open instincts, as Dewey pointed out, allow flexibility in behavior patterns, an extremely necessary skill for adaptable creatures. Open instincts, however, do not mean that they are necessarily weaker. Midgley argues that the fact that any set of open instincts needs to help the individual negotiate a host of unknown situations, means that they need to offer more guidance to the individual, not less:

> The less firmly the next action is settled in advance, the stronger must be the general desire that will lead to discovering it. More obviously still, mammals could not improve on the automatic brood-tending of bees merely by being more intelligent about what benefits infants. They have to *want* to benefit them. And they must want it more, not less, than bees, because they are so much freer, and could easily desert their infants if they had a mind to, which is the sort of thing that could simply never occur to a bee.[39]

Only a creature which knows what it is looking for and why is going to be able to survive in an unknown and changing environment.[40]

This emphasis on open, general tendencies, rather than closed, specific ones, creates a central place for the concept of learning. Individuals need to match their tendencies with the environment around them, in order to appropriately utilize their innate abilities. For intelligent, social beings, such as we are, this means learning how to utilize the environment for our needs. Once strategies of adaptation were learned, they were passed down through culture to the next generation. Extending Darwin's original insight, Midgley argues:

> On Darwin's suggestion, the relation of the natural social motives to morality would be much like the relation of natural curiosity to mathematics and science, or the relation of natural wonder and admiration to art, or that of natural amusability to jokes. These natural motives do not of themselves create the arts and institutions that channel them. But they provide a certain appropriate motivational force that is necessary to create these channels, and they also determine, sometimes in surprising ways, the direction which that force will take.[41]

For example:

> ... if we ask ... about our aesthetic sensibilities, it seems clear that these too, must have an inherited physical basis. These capacities, however, are not a clog or a fetter on our aesthetic life, instead, they are the basic apparatus which makes it possible. (... Bad luck; no natural susceptibility, no Beethoven.)[42]

Like Kropotkin and Dewey, Midgley sees culture as emerging from our evolutionary nature, restrained by its structure, but also extending and navigating its possibilities. Culture is not determined by nature, as Spencer argued, but neither is it independent of it. Nature is the raw material on which culture works, but the choice of cultural directions is motivated by our innate motives. Culture is continuous with nature, not socially constructed on the background of a blank slate.[43]

THE TELEOLOGICAL IMPLICATIONS OF HAVING NEEDS

In the Hobbesian version of our genetic makeup as presented by Midgley, our instincts are viewed as distinct entities operating independently of one another and in constant conflict. Our sense of sympathy conflicts with our aggression; our desire to be with people conflicts with our desire to be alone—a war of all against all.[44] This, of course, makes no evolutionary sense. A vision of warring instincts makes no sense when viewed on the background of evolution, in which individual animals needed to be coherent, functioning wholes, and not fragmented, conflicting beings that are incapable of acting in the world. Ethologists have been demonstrating that animals, stereotyped as acting randomly, with no apparent explanation for their behavior, are in fact mostly quite coherent wholes living structured and purposeful lives.[45] They are certainly not indiscriminate monsters as they were often portrayed in various animal myths:

> Recently, ethologists have taken the trouble to watch wolves systematically, between mealtimes, and have found them to be, by human standards, paragons of steadiness and good conduct. They pair for life, they are faithful and affectionate spouses and parents, they show great loyalty to their pack and great courage and persistence in the face of difficulties, they carefully respect one another's territories, keep their dens extremely clean and seldom kill anything that they do not need for dinner. If they fight with another wolf, the encounter normally ends with a submission.[46]

In order for this kind of steadiness to develop, motives cannot be in hopeless conflict with one another.

Behaviors find their source in a variety of motives interacting with one another. The absence of one behavior has implications for the animal as a whole. One cannot isolate a particular behavior from the coherency of the animal. Eibl-Eibesfeldt, quoted by Midgley, gives several examples of ways in which motives are interrelated in the building of behavior:

> While there is no friendship without aggression, there is also, with few exceptions, no friendship without parental care. . . . Love is not primarily a child of aggression, but has arisen with the evolution of parental care. . . . Among the animals that do not look after their young, we know of no group defence and no fighting partnerships. . . . Brood care, on the other hand, calls very early for individual partnerships and individualized cherishing of the young, and thereby offers the necessary basis for a differentiated social life. Aggression plays only a secondary role in strengthening a bond. The sexual drive, on the other hand, is extremely rarely used as a means of cohesion, although in the case of human beings it plays an important role in this respect. *The roots of love are not in sexuality*, although love makes use of it for the secondary strengthening of the bond.[47]

This point was particularly relevant to Lorenz's argument in his book *On Aggression*, in which he argues that aggression was not some evil instinct motivating indiscriminate killing in human society, but rather an indispensable aspect of personality.

> Here Lorenz is very much interested in the *value* of aggression, in the relation of pugnacity to vigorous effort, in people who "fight unremittingly" on behalf of the truth, or to defend the helpless, in the struggle for reform and the battle against evil generally. Saying this is only suggesting a field for study. But it ought to make us wary of people such as Arthur Koestler who say that aggression is a disease and ought to be chemically treated by pills or the like. Nobody knows how much of human life might go with it if that were tried.[48]

Midgley's argument is not only that each motivation is linked to the whole person, but also that removing one damages the entire system, including other aspects of our personality that we value, an argument we have seen Dewey make, as well. In addition, each and every one of our motives is part of what it means to be human, and their manifestations, therefore, are in a fundamental sense human needs. Relating to any of them as if they could possibly be bad for us, something fundamentally evil, makes no sense. They are, after all, a central part of who we are. Lorenz tried to show that aggression is a fundamental human need, and not an evil which needs to be exorcised from our make-up. So it is with all of our motives. Attempts by some sociobiologists to describe particular discrete behaviors as biologically proscribed, such as rape, strikes Midgley as radically reductionist, as they use an atomist model where each innate motive translates seamlessly into discrete actions.[49]

The important point here is that these innate motives form the set of goods for our lives. Our fundamental repertoire of motives, expanded by culture, is inherited.[50] Their expression are ends-in-themselves, they are fundamental to what it means to be a human being, and they should not be understood as means for the pursuit of some other end.

> Why affection? Why time-consuming greeting procedures, mutual grooming, dominance and submission displays, territorial boasting, and ritual conflict? Why play? Why (on the human scene) so much time spent in nonproductive communication of every kind—idle chatter, lovemaking, sport, laughter, song, dance, and storytelling, quarrels, ceremonial, mourning and weeping? Intelligence alone would not generate these ends. It would just calculate means. But these things are done for their own sake; they are a part of the activity that goes to make up the life proper to each species.[51]

There is no one end for a human life.[52] As Midgley says, how many ends there are is an empirical question, based on our psychological makeup.[53] Intelligence must then negotiate the relationship between the multiple and at times conflicting ends of a human life, as is discussed later.

The radical repression of any of our needs is, therefore, a retarding of our humanity. A life lived in isolation from others, deprived of sociability, is in a fundamental way a life unfulfilled, as is a life lived without love, laughter, play or curiosity. Taking animals out of their natural habitats, for example, and, as was the case in early zookeeping, having them live unnatural lives, without access to the goods that gave their life structure and meaning, led to depressed and dysfunctional animals. Even the repression of one motive could lead to general dysfunction.[54]

Midgley is consciously following Aristotle in asserting that biology is a central component of moral and philosophical discourse.[55] It is a teleological argument in that it argues that human beings are designed for a particular kind of life. For example, Midgley writes that "each creature has its own faculties and not others. And as Aristotle pointed out, for each creature it must be good to use its faculties and bad to be prevented from doing so."[56] The designing process is not necessarily a purposeful one, however, since the designing by natural selection and evolution is contingent and seemingly random, as opposed to what is often suggested by design, which is a purposeful designer. I will call it a proximate, rather than an ultimate, teleology, rooted in a discussion of proximate, rather than ultimate, ends. But the result has nevertheless been a life with needs internal to its structure that constitute the ends of a human life. Dewey's philosophy hinted at such a direction, by rejecting ultimate purposes, while suggesting proximate ones, but Midgley has fully elaborated what a proximate teleology looks like.

Each species is unique in its genetic makeup, and therefore has a distinctive motivational and behavioral makeup, just as it has distinctive physiological attributes.[57] Animal trainers know this, and they attempt to understand the needs of other animals, as a prerequisite to bridging the species gap:

> People who succeed well with them do not do so just by some abstract, magical human superiority, but by interacting socially with them—by attending to them and coming to understand how various things appear from each animal's point of view. To ignore or disbelieve in the existence of that point of view would be fatal to the attempt. The traditional assumption behind the domestication of animals has been that, as Thomas Nagel has put it, there is something which it is to be a bat, and similarly there is something which it is to be a horse or donkey, and to be this horse or donkey.[58]

To prove her point about human innate emotions forming human needs, and not rationality, Midgley uses an imaginary example, positing a hypothetically intelligent species whose biology is different from our own:

> The virtuous and super-intelligent Quongs are offering to adopt human babies. Shall we let them? What do we need to know first? The first thing, I should guess, concerns emotional communication. Do the Quongs smile and laugh? Do they understand smiles and laughter? Do they cry or understand crying? Do they ever lose their temper? Does speech—or its equivalent—among them play the same sort of emotional part that it does in human life—for instance, do they greet, thank, scold, swear, converse, tell stories? How much time do they give to their own children? Then—what about play? Do they play with their young at all? If so, how? Then, what are their gender arrangements—meaning, of course, not just sexual activity, but the division into roles of the two (or more) participating genders, throughout life? What singing, dancing or other such activities have they? What meaning do they attach to such words as love? Without going any further,

it seems clear that, unless they are the usual cheap substitute for alien beings which appears in films—that is, more or less people in make-up—we shall find that the answers to these questions give us some reasons to refuse their offer completely, even if reluctantly. . . . A human being needs a human life.[59]

These needs, or wants, or goods, form the pluralistic ends of a human life, and can play a pivotal role in rebuilding teleological ends for education, as is discussed in the final chapter.[60]

Human beings, therefore, share a common nature, which forms the substrate on which meaningful human life is based. Attempts to deny that humanity, to place upon it a life for which it is unsuited, we call dehumanization. The very idea of dehumanization is predicated on the idea that there is a human essence which has, in some fundamental sense, been degraded. Restoring our humanity presupposes that there is some essential humanity which needs to be restored.[61] Furthermore, according to Midgley, that essence is where humanity finds resources with which to resist socialization.

. . . the whole notion of natural human tastes which rulers must not distort or ignore—a notion on which we all rest when we resist bad institutions as "inhuman"—presupposes a firm biological basis in inherited human nature. Marx's central notion of "dehumanization" rests on this plinth, and his attacks on the notion of human nature are simply aimed at inadequate forms of it. If there were no such thing as human nature, the objections to a *Brave New World* existence could never arise. Conditioning is the tool of tyrants. Natural, inborn human spontaneity, seeking a more satisfying life even among people who have been brutally conditioned to know nothing but slavery, is the source of resistance to tyranny.[62]

Inhuman conditioning is resisted by the innate motives of human beings, which, serving as a precultural basis of human beings, can serve as a resource from which dehumanization can be intuitively understood and challenged. This point, too, is explored in the final chapter, when I discuss educational philosophy.

Midgley, therefore, is arguing that humans have a nature which is prescriptive for how to live their lives. It is a teleological argument in that it suggests that human life has purposes which are intrinsic to it. In order to see what the implications of such an argument are, I would like to turn to Midgley's discussion on feminism, which in many ways is a paradigmatic case for her argument, and a subject on which she coauthored a book.[63] It is not surprising that Midgley holds the controversial position that there are significant biological differences between men and women, both physiologically and psychologically/motivationally, and that these differences should matter.

FEMINISM AND HUMAN NATURE: A CASE STUDY IN TELEOLOGICAL THINKING

Although Midgley posits a human nature common to all humanity, she also argues that there are significant differences in innate nature between individuals, not only because their environments are different, but also because their genetic propensities are different. The notion of an individual with an innate, biological nature is a critical concept for Midgley. This nature, as I have shown, seems to resist the indiscriminate conditioning of society. It defines the intrinsic good of a life, and lies at the root of what it means to be human. "Bewick," Midgley notes, "was a *born* draughtsman, just as Ramanujan was a born mathematician. Such people exist."[64] If we had no ends, no purposes, there would be nothing to be free in order to do. For Midgley, this is in fact a central component of the liberal view of freedom, articulated, for example, by John Stuart Mill.[65] Because each one of us is biologically unique, any attempt by society to stamp us into one model denies us our humanity. Society and culture have oppressed humanity, by forcing us to conform to societal norms, and by not allowing us to spontaneously grow as our natures demand. Society needs to be curtailed in order to allow the true individual to flourish:

> What then, is the central reason why we should attend more to the claims of individuals, and less to those of society? . . . [Mill] takes the point of view of suppressed individuals themselves, describing the miserable frustration of their lot and celebrating, by contrast, the splendour of the free, spontaneous activity that ought to replace it. Here, Mill repeatedly uses strong metaphors from nature. The suppressed individual is (he says) like a tree spoiled by pollarding or a bush clipped down into fancy shapes by a gardener. More sharply still, he remarks that Society's "ideal of character is to be without any marked character; to maim by compression, *like a Chinese lady's foot,* every part of human nature which stands out prominently, and tends to make the person markedly dissimilar in outline to commonplace humanity. These strong images help to build up the general value-judgment that

free actions and free individuals are in themselves simply much better things—far more precious—than conditioned and automatic ones.[66]

Some feminists have argued that the difference between the sexes is in fact a difference between individuals. While women may be inclined to certain behaviors that have been considered traditionally female, there are many women whose natural tendencies lead them to behaviors that are traditionally male, as well as men who have tendencies leading them to behaviors that are traditionally female. While accepting this, Midgley probably believes that the two bell curves are not identical, but rather distinct, although overlapping.[67] Just as there are differences among individuals, so too there are differences between the sexes. For Midgley, a view of men and women as identical is another attempt by society to stifle our unique nature.

Social Darwinists of the nineteenth century also argued that there were differences between the sexes, with disastrous consequences.[68] One side was clearly seen as superior to the other. Many feminists, choosing a strategy similar to those who argued against innate differences among the races (a prominent theme in the first-generation Darwinian debates), have subsequently argued that there are no categorical differences between groups of human beings, and that men and women are essentially the same—different from one another in the same way that two men or women are different from each other. Although there were physical differences in men's and women's biology, they certainly didn't manifest themselves as differences in character. Any differences between men and women beyond the most obviously physical ones were cultural, not biological.[69]

Midgley obviously finds this to be an untenable position, the once-again linking of a leftist, progressive political agenda with a blank-slate theory of human nature. Applying Darwin's theories, she argues that the distinction between male and female is one of central significance to the respective development of males and females in a species. Physiological differences and evolutionary roles that are so distinct from one another will by necessity lead, through natural selection, to different proximate strategies. Ethologists are able to identify different behaviors between males and females in all species. It is inconceivable that such a paramount difference in the natural world would not apply to human beings as well.[70] Midgley tries to disconnect equality from a disembodied sameness, and instead advocate for equality in its embodied context:

> ... equality is not sameness. A belief in sameness here is both irrelevant to the struggle for equal rights and inconsistent with the facts. It

> ignores massive evidence of sex differences in brain and nerve structure occurring long before birth, and also of behavioural differences which are evidently independent of culture and sometimes contrary to it. It amounts to an extraordinary abstract notion—evidently held on moral grounds—of the original human being as something neutral, sexless and indeterminate, something wholly detached from the brain and nervous system.[71]

For many feminists, however, the body was something which was seen as the "ball and chain" which prevented equal treatment. Women give birth to babies and they breastfeed; therefore women are castigated into the role of mothers and caretakers, doomed to a life at home away from the real arena of human life. Midgley views the existentialist Simone De Beauvoir as an exemplar of such an approach:

> Simone de Beauvoir, among many others, declares that there is something not just frightening, but metaphysically degrading, about pregnancy and childbirth. A pregnant woman is, she says, "alienated"; in her, the species is taking over the individual; she is "in the iron grip of the species." Both the father and the child are violating her sacred individuality.[72]

So too, Shulamith Firestone:

> Thus Shulamith Firestone writes that 'women throughout history before birth control were at the continual mercy of their biology'; and that the substitution of *in vitro* pregnancy for current methods would 'free women from their biology.' What this complaint really seems to express is a horror of the body—especially, of course, of childbirth—as a threat to the free mind and soul.[73]

Many feminists, therefore, have tried to give equal treatment to women by ignoring what makes them different as an unfortunate and even alienating, although isolated and detached phenomenon. Women are the same as men, save for their biological burden, which needs to be somehow overcome so that they can be men's true equals. Huxley held such a position.

Midgley uses examples from feminist-supported legislation to show the logical absurdity of this "equality as sameness" strategy. In arguments over the application of equal treatment under American law, for example, legislators chose to pursue feminist legislation through the "equal as sameness" argument, and therefore defined discrimination as any case where one group is given benefits and another group is denied the same benefits. This became a problem when relating to pregnancy, for instance, which was obviously applicable to only one group. In order to maintain the "equality as sameness" argument, however, women who were to get maternity benefits could not justify getting them because they were women, but rather because they happened to be the ones getting pregnant. If men were to get pregnant, they would also receive benefits.[74] This kind of gender-neutral relationship to pregnancy, ignoring by definition the fact that women are the only ones getting pregnant, could, and occasionally did, backfire in practice. Midgley and Hughes cite a 1974 U.S. Supreme Court case, in which a California insurance company was found not to be in breach of the Fourteenth Amendment, which guarantees equal protection under the law, when it allowed payments for a wide range of disabilities, but refused disability payments for pregnancy and childbirth. The Supreme Court's claim was that there was no discrimination involved since men were also not given disability payments for pregnancy! No one was receiving pregnancy as a disability—men or women. The "equality as sameness" argument had not allowed the law to express the special demands of a distinct group owing to their different and distinct needs.[75] Midgley and Hughes present rape legislation, and the absurd attempt to make it gender-neutral, as another case study.[76]

The biological difference of pregnancy, however, is a difference of physiology, and not of motives. It might be the exceptional difference, therefore, which proves the rule. One can accept the fact that there are physiological differences without accepting the fact that there are personality differences as well. Midgley cites that it is often claimed that "male" and "female" characteristics can be found in both men and women, and their manifestations are a question of socialization.[77] If men in our society are more analytic and women more emotional, it is a question of socialization, and both sides are capable of benefiting from the other set of skills.

Midgley, of course, can't accept a barrier between the physical and the psychological. Mind and body are contiguous with one another, and, similar to the rest of the animal world, differences in body accompany differences in psychological makeup. She recognizes that the wrong set of differences have unfairly been associated with static sex roles, but she argues that we need to get the differences right, rather than brazenly abandon the notion of differences.[78] In any event, society won't allow us to abandon the differences, nor should it. Like exorcising aggression, ignoring sex differences can cause significant damage to the person as a whole, so we had better work on getting them right, rather than ignoring them and denying their existence.[79] There are differences in men's and women's character. For example, it is a universal fact that in cultures throughout the world, boys and girls play separately. Why would this happen if there weren't differences?

> Those devoted to excluding innate causes treat this as a mere reflection of the adult division into sex roles. But children do not reflect everything which goes on around them, and anyway that division itself remains to be explained. If the sexes are psychologically indistinguishable, why is it universal, which it certainly is? It would be very strange if these groupings did not reflect a natural difference. The important thing is to avoid loading that real difference with irrelevant and dangerous symbolism.[80]

Women and men are different, and therefore have different needs. This isn't to say that they are radically different, or that there aren't differences that are accentuated by culture, or differences that are the result of culture, but these are empirical questions which need to be researched.[81] Midgley, for example, hypothesizes that the stereotyped idea that women are emotional, intuitive, and nurturing, while men are analytic and abstract thinkers has both a factual basis, emerging from differences in their biology, but also from the interplay between biology and society, and obviously at times solely from socialization, as well. Midgley theorizes that women, as the primary caretakers of infants as a result of the need to breastfeed, and psychologically through innate motives for bonding with the child which fortifies that role, develop the ability to interpret behavior and communication of an infant who is preverbal. That skill continues to develop in a myriad of social relations, as the sociological literature on the ways women think and communicate increasingly describes.[82] Men, on the other hand, who were not as biologically essential for childrearing, and were therefore less likely to evolutionarily develop instincts and motives adapted to fortify such a role, were largely the ones who received training in analytic argument, often with the goal of winning the argument (but not always for clarifying the truth), through their professional training, for example, for the church, politics, and law.[83] The results of these differing paths of motives and socialization lead to the often stereotyped view of men as analytical and women as intuitive. It is not that men and women don't both have the ability to either "intuitively" or "analytically" reason. It is not only that they have had opportunities to develop only one particular aspect of reasoning. They also have different motives which support the development of one type of reasoning over another.[84]

If one must choose between the two stereotypes, it is clear that Midgley prefers the female one. On the one hand, there is nothing wrong with the female's way of knowing, which is part of a network of connectedness to others.[85] The male stereotype, on the other hand, continues the atomized, disembodied ideal of separation between mind and body, and reason and emotion. The "isolated, individual, competitive [male] individual, each hell-bent on his own interest" is in fact reliant on the female stereotype in order to operate. Someone needs to worry about the children's well-being.[86] Even if they were interested in pursuing the male ideal, women are led back by the biology of childcare with its nurturing demands, to the network of connections and obligations that make up a human life.[87] It is no wonder that those feminists who chose the male ideal were alienated from their biology, which reminded them that life is not an atomized existence. Midgley's strategy is to remove the ideal of the atomized individual—made possible only by women's subjugation to that ideal—and to build a more realistic view of life for both sexes, and for society as a whole:

> We have come to the end of the road for sex-linked individualism. It has been rumbled. The idea that every man should properly look out only for himself, while every woman should stand staunchly and obediently behind him as he does so, is not sensible and will not wash once attention has been drawn to it. We have a choice. We can either extend the individualism which has been almost a religion in the West since the eighteenth century consistently to *both* sexes, or we can admit its limitations, treat it with much more caution, and put it in its place as only one element in a more realistic attitude to life for everybody.[88]

Feminists have traditionally recoiled from this strategy which Midgley is proposing, since differences between the sexes can be a slippery slope, and have notoriously been interpreted as suggesting the inferiority of women. Midgley recognizes the historical reasons of arguing from sameness, but nevertheless argues that the opposite should be the strategy. Society should relate to the sexes according to their motives and needs—when they are identical, society should relate to the sexes identically; when they are different, society must relate differently in order to insure true equality.[89] Whether, in fact, essential differences between the sexes can be identified, without reading into them the value biases that have led to false and often dangerous dichotomies, is obviously a contentious issue, but Midgley's point, where claiming that there are no differences is also a value bias with dangerous implications, strikes me as sound.

Up until this point, I have shown that Midgley's teleology emerges from her claim that there is a plurality of motives or needs which make up the aims of human life. I have not yet shown, however, how a plurality of needs is then to be negotiated into a single human life. Are there needs that are more important than others? Do all needs have to find equal expression, or can one need dominate others? Are needs fixed, or can culture shape, or even form, different sets of needs? By what process and by what criteria are such priorities developed? In many ways it was the same question that occupied Dewey's response to Huxley's *Evolution and Ethics*: Does human nature proscribe a strong and unified good out of the multiple goods of human motives? It is to the question of negotiating between different needs in building a human life that I now turn.

ON BUILDING A WHOLE LIFE

Midgley claims, as we have seen, that there is a plurality of motives in human life, and that all are, in a weak sense, essential to human life. It is in a weak sense because, if left at this point, the argument seems to build a set of distinct aims which are incommensurate with one another, and therefore a list with no way of prioritizing or choosing. Hume argued such a position, claiming that since the moral sentiments are rooted in our passions, they cannot be prioritized, since prioritization is a rational activity independent of our passions.[90] They live in separate realms. Midgley, as we shall see, rejects this as just another one of the dichotomies which have left humanity disembodied, and instead argues that rationality in fact evolved, as part of our social natures, and as part of a strategy of making sense of our conflicting motives.

For Midgley, culture is rather the historically compiled knowledge of a group of human beings' struggles to build a vision of the good life, along with a way of achieving it. Natural selection is not a perfecting mechanism, but rather a maximizing mechanism, making creatures "good enough" to do what they do: "it seems to be a process which makes it inevitable that our natural desires must conflict."[91] As Midgley argues:

> We want incompatible things, and want them badly. We are fairly aggressive, yet we want company and depend on long-term enterprises. We love those around and need their love, yet we want independence and need to wander. We are restlessly curious and meddling, yet long for permanence.[92]

Our "constant ambivalence" as human beings stems from the fact that these conflicting needs are real, and that choosing one direction over another really does mean sacrificing one part of ourselves for another part.[93]

Midgley argues, however, that while we have multiple needs in our lives and therefore ends, these needs are not mutually exclusive of one another, fighting for control over our identities.[94] Because we are distinct individuals, who must function as an emotional and rational whole, a schizophrenic mass of conflicting and mutually irreconcilable desires is simply not an option. Each of us attempts to build a whole person, who can function as a whole and who is a coherent, comprehensible individual. Our various motives must be integrated into a sensible identity. This is not simply a choice that we make. We are, like all animals, a unified whole.[95] Our consciousness aids us to act as a whole, as shall be discussed, but it does not create the desire. Holding oneself together is an innate human desire, central to being a functioning living unit.[96]

Our motives, therefore, are a network of goods which the individual must negotiate in order to build a coherent life. Individuals will choose to prioritize and develop their motives differently, based on their relationships to the whole, and yet all of the parts are in some measure important for the proper functioning of the whole. Aggression, for example, even if it were seen as something evil, which it is not, still cannot be surgically eliminated without affecting the functioning of the whole. And since every motivation is an end in itself, critical to the proper functioning of the whole, each part is necessary not only for its own sake, but also for the sake of the whole. No one part of the system should so dominate the individual that the rest of the individual's natural desires are subjugated. Midgley argues that when such a takeover takes place, a hemorrhaging, so to speak, of one part of the whole occurs.[97] Such a hemorrhaging can be called unnatural, in the same way that cancer, in one sense a natural occurrence, can be considered unnatural in that it is disastrous for the proper functioning of the whole:

> But redirected aggression and so on can properly be called *unnatural* when we think of nature in the fuller sense, not just as an assembly of parts, but as an organized whole. They are parts which will ruin the shape of that whole if they are allowed in any sense to take over.[98]

Here Midgley argues that a good can turn into a bad when it begins to dominate the proper functioning of the whole. Echoing Dewey, she is arguing that nature is a whole, and it is the shape of the whole which allows us to evaluate the worth of the parts. Any part of the whole which undermines it, while still natural in the weak sense, is unnatural in the strong sense, like the weed in the garden. Any part of one's innate nature is good in a weak sense, but only our nature as a whole can be conceived of as a good. Elsewhere Midgley argues that it would be best to consider our natures as neither good nor bad, but rather as the raw material from which we choose.[99] Good or evil is a result of the choices we make in the caring and development of the garden as a whole.

> I have suggested that, when we wonder whether something is good, common sense will naturally direct our attention to *wants*. And it is because our wants conflict that we have problems. We often need a priority system by which we can say, "both are good in different ways, but this good matters more than that one." So the full facts of conflict are essential to understanding the meaning of terms like *good*. Thus *these* facts are not irrelevant to values. If we say something is good or bad for human beings, we must take our species' actual needs and wants as facts, as something given. And the same would be true if we were speaking of any other species. It is hard to see what would be meant by calling good something that is not in any way wanted or needed by a living creature. Should we therefore say that everything we want is good? In a minimal sense this is right; everything we want has to have something good about it, otherwise we could not want it. But of course we cannot stop there, or *good* would simply mean *wanted*. We must go further because of conflict, because of the clash and competition among our various wants. What is good in a stronger, more considered sense must be wanted, not just by someone's casual impulse, but by *him as a whole* and on grounds conveyable to those around him.[100]

Because we function as a whole, we can also predict behavior, expect certain behavior from certain species, and indeed, from certain individuals. When someone is not acting like him/herself, we mean that we have come to expect a certain whole, a certain priority system, which this behavior does not seem to reflect.[101]

Midgley's argument consciously follows Aristotle's idea of the golden mean.[102] We have natural motives that in their right measure are critical aspects of being human. Too little or too much of any motivation distorts the human life as a whole:

> As Aristotle suggested, we need a mean. Fear is all right in moderation. There should be neither too much nor too little of it. And this moderate level is not just an arithmetical mean, halfway between extreme

> rashness and extreme cowardice. It involves fearing the *right* things, not the wrong ones, and fearing them as much as, not more than, their nature calls for.[103]

This kind of balanced answer is in fact much more substantial and less evasive than it looks. It would be evasive if it did not suggest a context for deciding on "the right level." But it does. It refers us to the context of a whole human life, and sends us for the details to investigate how the various parts of that life are lived and how they fit together.[104]

The investigation of "how the various parts of that life are lived and how they fit together" is the job of intelligence.[105] In this, humans are different from other animals who have conflicts between their motives—for example, Darwin's observation of the conflicting desires among migratory birds between taking care of their offspring and their stronger immediate desires, at the required time, to abandon everything and migrate. Only humans, however, have such an advanced intelligence, which allows them to be consciously aware of conflicting desires, and to use that awareness to negotiate the conflicts between motives, and to actively choose a path of conduct.[106] Notice how similar this description is to Dewey's description of the building of an individual life through the rational negotiation between conflicting motives.

The fact that human beings are a social species meant that their intelligence would be rooted in their social context. Since it is our sociability that forms the central part of our emotional structure, it makes sense that our reason would be motivated to protect and strengthen our social sympathies.[107] Human beings' intelligence is directed, naturally, at being a functioning whole, part of the world which surrounds it. But because we have conflicting motives, the desire for wholeness is often undermined by stronger, yet less social desires. The passion of the moment, for example, can lead to anger overtaking all other motives momentarily, which is often observed among other primates as well.[108] What is different about humans is that because of consciousness and, therefore, memory, the human being, unlike its primate cousins, remembers what happens and regrets its occurrence.[109] Why do people regret actions? Midgley argues, again following Darwin's lead, that once the passion of any particular emotion has abated, the more general, but no less natural desire to live a balanced life in companionship with others leads to regret at not having inhibited one's actions.

> Let us consider the predicament of primitive man. He is not without natural inhibitions, but his inhibitions are weak. He cannot, like the

dove or the roe deer, cheerfully mince up his family in cold blood and without provocation. (If he could, he would certainly not have survived long after the invention of weapons, nor could the prolonged demanding helplessness of human infants ever have been tolerated.) He has a certain natural dislike for such activities, but it is weak and often overborne. He does horrible things and is filled with remorse afterwards. These conflicts are prerational; they do not fall between his reason and his primitive motives, but between two groups of those primitive motives themselves. They are not the result of thinking; more likely they are among the things that first made him think. They are not the result of social conditioning; they are part of its cause. Intelligence is evolved as a way of dealing with puzzles, an alternative to the strength that can kick its way past them or the inertness that can hide from them. . . . I would add that only a creature of this intermediate kind, with inhibitions that are weak *but genuine*, would ever have been likely to develop a morality. Conceptual thought formalizes and extends what instinct started.[110]

Morality emerges from our moral sentiment and is extended through our rationality, in the need for both society and each one of us to build a conceptual map with which to choose between conflicting motives. When should I react aggressively and when should I not? When should I forgive rather than continue to be angry or hurt? Nor is morality, in Midgley's view, limited to how I relate to others. Choosing between being an artist and an athlete is a moral choice, in that it is choosing between conflicting aspects of who I am, and deciding who I want to be. For Midgley, the emotional and rational prioritizing of different motives is what morality is about.[111]

No individual is alone in negotiating these choices. For Midgley, as for Dewey, culture is a product of the collective work of human beings of a particular group negotiating the various conflicts of the human species, and offering a strategy for its resolution:

What, then, does intelligence do? It helps us to build a culture—a set of customs expressing a priority system, which will show us how to settle these conflicts in certain agreed ways so as to make the stresses of decision more or less bearable. Within that general framework, it then helps us to decide, by creative effort, how to settle the further clashes that constantly arise and that our culture has not settled. It is because there is no pre-set, universal priority system available that cultures differ so much. Yet it is because their basic problems are still the same that cultures are, none the less, so similar.[112]

Intelligence is the tool for constructing our open, interlocking, yet also conflicting motives into a sensible, functioning whole.[113] This task, by definition, is never ending, but culture has allowed us to benefit from past negotiations of the meaning of the human condition.

In this view, the historical dichotomy between reason and emotions breaks down, as reason itself is rooted in emotion, and functions in connection with our emotional lives.[114] Our emotions, like those of other animals, have a certain structure which suggests the way in which conflicts between motives are to be resolved. For humans, that structure is rooted primarily in our sociability, that guides us in the direction within which reason works. Without it, reason would be lost.[115] It would be a capacity without a direction in which to function.

Developing human intelligence disconnected from social motives, which are its proper context, can be a dangerous cultural experiment. Without the proper ends as its goal, intelligence can be applied to the irrational promotion of any number of sordid ends.[116] Our species has a structure which directs our actions, however, and therefore our intelligence is not lost in the universe without bearings. It is our emotional motives which give structure and purpose to our intellectual capacities. Commenting on Kant's focus on reason, as opposed to emotions, as the center of our pursuit of human ends, Midgley comments:

> And someone with an undeveloped heart—someone who simply doesn't care very much what happens to other people—won't be moved by information about it, and probably won't pursue the enquiry. Kant, when he made morality essentially a matter of reason, took for granted an emotional background which he did not notice. Strong sympathy for other people, and indeed a positive passion for justice, are necessary factors if "practical reason" is to move mountains in the way he wanted.[117]

Moral judgment, therefore, like rationality itself, should not be developed independently of our emotional concerns, but rather based on them. We can trust our instincts as legitimate guides to right actions. The emotivists, Midgley asserts, were right about the central place of emotions in our moral life; they were wrong in disconnecting them from our reason. Reason completes the work of our emotions.[118]

Midgley is advocating a view of morality where what one feels is an essential part of the moral universe. As Darwin argued, one cannot be truly moral if one's emotional life is disconnected from one's actions. Rational support of equal treatment of blacks, women or homosexuals, while emotionally seeing them as something inferior, or as a sex object, or as something repulsive, is building a schizophrenic view of self, where the inner life either doesn't matter, or is nobody else's business. Once accepting that the two worlds are part of the same one, that emotions and reason are connected, however, their separation is not an option. Our reason completes the work of our emotions. Our moral intuitions, however, are reason's starting point.

In the long run, a divorce between motive and action is disastrous. Sympathy is not just an act, but a character trait, as pointed out by Darwin. Those that act kindly but are motivated by duty, rather than kindness, will inevitably be incapable of acting truly kindly. Their actions will be hollow, and will be perceived as such.

> It really is far better to do what one ought *willingly*—feeling the need for it, seeing the point of it and throwing one's weight behind it—than to do it from the kind of "duty" that is totally reluctant and alienated, as is conveyed in the depressing phrase "a duty visit." If duty is merely the voice of the community—if nothing in ourselves accepts it, if our feelings do not respond to it—then, however good conduct may be, it is dead, and something absolutely central to morals is lost.[119]

The inner life and the outer life cannot be divorced, but are inseparable components of the same individual who strives to work as a whole.

Emotions and thought, therefore, are both contained in our actions. Emotions affect our moral reasoning by giving it direction, a framework to complete. But thought also can affect our motives. As has been shown, all motives, in a weak sense, are goods, and only become strong goods through the process of integration into a whole. Their integration is through a process of reasoning which then feeds back into the structure of the motives themselves. As Midgley argues, again echoing Dewey: "thought, feeling and action are conceptually, not contingently, connected. They are aspects of the one thing: conduct." [120] Emotions motivate thought, and thought structures our emotions. Together they make up our actions, which become, over time, part of our emotional selves:

> "Good will" is not the power to do the right thing suddenly while still wallowing in habitual malice, envy, self-pity or the fear of life.

> It is the power to change such emotional habits, over time, through vigorous attending and imagining, into better ones, which will incidentally be ones from which doing the right thing becomes natural. Certainly occasional actions ahead of and contrary to one's current state of feeling are possible and necessary. As Aristotle says, someone who is trying to become just must start by doing just acts, without enjoying them. Only after practising in this way for some time will he begin to enjoy them. Only after practising in this way for some time will he begin to enjoy them as the just man does. . . . Normally, good will is utterly dependent on self-knowledge, on studying and understanding our disastrous habits of feeling. By attending to such habits *and* attending to considerations which show that they stink, we can in fact gradually control them, replacing them by better ones and feeling differently. There is nothing artificial, inhuman or bogus about this. The artificiality belongs to the other enterprise—that of trying to act correctly without attention to one's feelings.[121]

Our knowledge is embodied, part of our bodies as well as our minds. It needs to "get under our skin." Knowledge that rests only as part of the intellect, disconnected from our emotional selves, remains segregated from who we really are. Through building correct habits over time we can internalize these habits through our very physicality into our emotional makeup. As Dewey argued, we can only know what it means to stand straight, and to learn how to stand straight, by experiencing and practicing standing straight.[122] We do not truly know something until it is integrated into our bodies and being.[123]

Our wholeness, therefore, is not to be assumed. While we are born with a nature which offers us guidance in negotiating choices, and we are born into a culture which has done much work for us (but also which needs to be reexamined constantly), we are nonetheless constantly reshaping who we are as we choose what aspects of our personality we should allow to have prominence. Our next set of choices will emerge from the new starting point which has been formed. Here our freedom is clear. Each individual chooses a different path from which to develop the raw material which is shared with others. The choices that are made determine who the person is. The weak ends offered by our innate nature are transformed into strong ends as we forge a life. That life is forged based on the raw material which each of us is given, within the context of the culture to which we are born. Because we are conscious beings, the building of that life is based on conscious thought, not as something separate from our motives and desires, but integrated within it. In a proper sense, Midgley argues, therefore, we can see life's purpose and life's meaning as attaining our full growth as human beings.[124] Like Aristotle, it is to become the tree from the acorn seed:

> Our unity as individuals is not something given. It is a continuing, lifelong project, an effort constantly undertaken in the face of endless disintegrating forces. We, as well as every mouse and every apple-tree, struggle for this wholeness as best we can throughout our lives, undiscouraged by endless obstacles. And we struggle in quite a different style from them because our struggle is conscious.[125]

We are unique among the animals in that our struggle to build a coherent whole takes place as conscious beings. We remember our past decisions. We have guilt. We cannot let go of what we have done, and we cannot help asking ourselves where we are going. Unlike the chimpanzee that has fought with a friend and quickly forgets the incident, we live in a world of memories and hopes which help us to consciously build a life which has a direction and an essence. Being a solid human being is in fact just that—having a core which is slowly formed from conflicting desires, which helps one to make future choices which strengthen, although at times also challenges, that core.[126]

MORAL OBJECTIVITY AND THE REALITY OF EVIL

Midgley's philosophy, like Kropotkin's and Dewey's, is, at its heart, optimistic about the human condition. Our innate motives are not, by nature, evil. This, of course, contradicts Huxley's views in the first generation, Dawkins in the second, and a growing scientific and popular Darwinian literature which places rape, murder and genocide as natural outcomes of particularly male biology.[127] Natural philosophers who sought to root human norms in innate human nature needed to explain how human nature was not the source of evil, and, if it was not, then what was its source. Darwin, as discussed, saw evil originating from a lack of motives, and Spencer, from the gap between our present innate nature and the nature which is adapted to the next cultural epoch. Midgley's naturalistic explanation of motives, suggesting that there is a way of choosing correctly between conflicting motives based on how they contribute to the whole, also suggests that there can be a wrong choice. If one can choose the good, one can also choose the bad. Examining how she looks at radical wrongdoing, what she calls wickedness, gives insight into how her moral system operates.

In modern times it has been tempting to explain away the idea of wickedness, as part of a general relativizing in culture that argues that values are relative, and therefore we have no basis from which to judge others. Midgley divides relativism into two: the relativism that prevents us from judging another culture, and the relativism that prevents us from judging another person's actions from within our culture, which she labels subjectivism.[128] Midgley, however, is arguing for a universal standard—in some basic way the fact that we are all human beings offers us a common standard. We can judge each other, and we can judge our culture and other cultures, although she is certainly aware that our judgments need to be the right ones, and that too often they are not. I return to this question later when examining how Midgley works from facts to values.

Part of the shrinking of judgment has been a sweeping fatalism that, in some fundamental ways, allows us to escape responsibility for our actions. One who kills another human being, for example, should not be blamed for his actions, but the political system which is responsible for the socioeconomic situation in which s/he was raised; or the broken home in which s/he was abused; or, in an example of the latest version of fatalism, the generous proportion of testoterone which made him do it should be blamed. All this suggests that the individual is a passive, and not an active, agent shaping his/her life, either because of his/her social experiences or because of his/her biology.[129] It suggests, in a mistake that Midgley claims that Aristotle at times made, that morality—conscious choice between alternatives as to the right thing to do—is superfluous, and health is synonymous with morals. Darwin, as I have pointed out, linked morality with mental health. If one is raised well, one shall live life the way that it was structured to be lived.[130] Those who act immorally are, in fact, mentally ill. For Midgley, this whole direction is an evasion of personal responsibility, and it ignores the choices that we make to become who we are. Instead of various levels of blame and punishment, it turns all wrongdoers into patients needing treatment, rather than potentially deliberate criminals.[131]

We are not angels. That means that we are not flawless beings without conflict. All of our traits have their value but also have their dangerous side. Aggression, as has been shown, is a central part of any human life, but it is dangerous if not watched and channeled correctly, as is jealousy or fear. Less suspected motives can be no less dangerous—love, loyalty, and curiosity. Each motive brings with it its light and dark side. Our impulses are not something alien to us, but are part and parcel of who we are. Denying their existence is to deny ourselves, and to reject our wholeness.

This view is very different from Huxley's and Dawkins's, who both see the natural state as one which is essentially self-centered and violent. For them, evolution proves that nature is dangerous. Midgley, aware that her philosophy is rather optimistic when it comes to the moral potential of human nature, nevertheless recognizes that moral behavior is not the only natural option. There is a natural history to evil and wickedness as well. But, as we have shown, describing any particular human motive as bad, or human nature in general as violent and destructive, is to reject the view that human life emerges from our sociability and our need to function as a whole. While destructive behavior can emerge, an overall destructive life strategy of self-interest at the expense of others simply doesn't make evolutionary sense.

For Midgley, therefore, evil is not the presence of a particular motive, but rather, as for Darwin and for Dewey, the absence of others. She argues that the urge to do wrong finds its root in important parts of our personality. There is nothing wrong with self-regard, for example, or the pursuit of honor. There is something wrong when these goals are pursued without a healthy regard for others to balance it. Without its balance, honor becomes obsessive and dangerous.[132] All motives have the danger of eclipsing the whole. And some of these motives are especially dangerous, particularly those that blind us to our responsibilities to the well-being of others. Being obsessive, therefore, threatens the overall balance of one's humanity:

> Obsession has to carry with it the atrophy and gradual death of all faculties not involved in whatever may be the obsessing occupation. And among these faculties is the power of caring for others, in so far as they are not the objects of obsession. To let an obsession take one over is therefore always to consent, in some degree, both to one's own death and to that of others. Or—to look at it another way—a destructive attitude to others, and to one's own nature, can be satisfied by cultivating an obsession.[133]

Midgley paraphrases Aristotle's observation that wrongdoing comes in two varieties.[134] There are those that do wrong because of a weak will. They know what the right action should be, but nevertheless surrender to their opposing motive. The action is done, but there is guilt and remorse involved. The other category, the one less dealt with by philosophers, according to Midgley, is doing wrong without remorse and without guilt. Here the actions are either not perceived as wrong at all, but as the right thing to do, or the actions are freed of any characterization of right or wrong.[135]

In the end, these categories are not in opposition, but rather options on a continuum.[136] Weakness is a manifestation of immoral behavior that can make remorseless behavior possible. Those who choose to ignore the greater moral landscape, and instead choose a selective mapping which supports their particular obsessions, have allowed laziness to slip into deceit. Midgley cites *Dr. Jekyll and Mr. Hyde* as a case in point. Jekyll, by weakness of spirit, has allowed Hyde to appear. He has amputated his dark side from his personal responsibility and set it free. Hyde, liberated from the weak, confused person called Jekyll, is cut loose, and roams free without remorse. Yet Hyde cannot appear without Jekyll's surrender of his moral unity to Hyde's existence. Jekyll allows Hyde to exist by refusing to take responsibility for his whole self.[137]

Taking responsibility for ourselves as a whole is harder in the modern world, as the implications of our actions extend out into space (to distant parts of the globe) and time (to future generations). This does not mean that we don't know what our actions do, but because they are so distant, we can effectively ignore them, and convince ourselves that we are not responsible. For example, while many people would join in protest over sweatshop conditions at their doorstep and among their neighbors, Westerners continue to support sweatshops and conditions tantamount to slave labor throughout the world through their consumption patterns. Alternatively, we continue to fill the atmosphere with greenhouse gases, increasingly aware of what the implications of such actions are to the planet and to our children. We choose to remove these actions from our moral universe.[138] Their distance from us allows us to choose such a strategy. As Midgley states:

> In the First World War, when a staff officer was eventually sent out to France to examine the battlefield, he broke down in tears at the sight of it, and exclaimed, "Have we really been ordering men to advance through all that mud?" This is a simple case of factual ignorance, flowing from negligence. Negligence on that scale however, is not excusable casualness. It is, as we would normally say, criminal. The general recipe for inexcusable acts is neither madness nor a bizarre morality, but a steady refusal to attend both to the consequences of one's actions and to the principles involved.[139]

The combination of modern bureaucracy and technology plays a large role in this escape from responsibility in our modern situation, a point made by Hannah Arendt in *Eichmann in Jerusalem,* when trying to come to grips with what she called "the banality of evil" when reporting on the Eichmann trial.[140] Quoting Arendt, Midgley argues: "'The essence of totalitarian government, and perhaps the nature of every bureaucracy, is to make functionaries and mere cogs in the administrative machinery out of men, and thus to dehumanize them."[141] This has obviously made it harder to see the implications of one's actions, and to take responsibility for them. But this does not absolve one of responsibility.

Wickedness, therefore, is fundamentally connected to dehumanization. It is not the flourishing of some primally wicked, destructive and sadistic impulses, whose origins would be impossible to explain from an evolutionary perspective, but rather particular motives overtaking the personality, out of balance with the human being as a whole. It is the loss of the whole human being which makes the wicked person, in a very basic way, inhuman. The loss of the whole person leads to the hollowing out of the center, and therefore to setting free motives without context. The person stops taking responsibility for his/her life as a whole. The individual allows the takeover to take place, and at times encourages the takeover to take place. Wickedness emerges in the world either by obsessive pursuit of one part of his/her personality at the expense of his/her humanity as a whole, or through active omission of responsibility for one's whole person.

BREAKING DOWN THE IS/OUGHT DICHOTOMY

Life's integration is the goal of a human life, and it affords Midgley the possibility of making moral judgments. When one part of the personality overtakes, when no balanced center emerges, the individual loses his/her moral compass, and evil emerges. Midgley's natural philosophy is the source of her ability to judge good and evil. We can judge one another, and we can judge ourselves. Quoting the moral philosopher Geoffrey Warnock, Midgley remarks:

> I believe that we all have, and should not let ourselves be bullied out of, the conviction that at least some questions as to what is good or bad for people, what is harmful or beneficial, are not in any serious sense matters of opinion. That it is a bad thing to be tortured or starved, humiliated or hurt, is not an opinion; it is a fact. That it is better for people to be loved and attended to, rather than hated or neglected, is again a plain fact, not a matter of opinion.[142]

Since the beginning of the twentieth century, however, moral philosophy had withdrawn from such judgments, allowing and often encouraging a vast relativism to take over. She traces the power of relativism in mid-twentieth century Western culture in general, and in philosophy in particular, to the work of G. E. Moore, whose philosophical writings at the beginning of the twentieth century were a strong reaction to the evolutionary ethics of Social Darwinism, particularly that of Herbert Spencer.

Moore famously elaborated on the is/ought dichotomy first articulated by David Hume, where Hume shows the philosophical problem of moving from a factual statement about the world to a value statement about it, what Moore labeled the naturalistic fallacy. In Hume's words:

> In every system of morality, which I have hitherto met with, I have always remark'd, that the author proceeds for some time in the ordinary way of reasoning, and establishes the being of a God, or makes observations concerning human affairs; when of a sudden I am surpiz'd to find, that instead of the usual copulations of propositions, *is*, and *is not*, I meet with no proposition that is not connected with an *ought*, or an *ought not*. This change is imperceptible; but is, however, of the last consequence. For as this *ought*, or *ought not*, expresses some new relation or affirmation, 'tis necessary that it shou'd be observ'd and explain'd; and at the same time that a reason should be given, for what seems altogether inconceivable, how this new relation can be a deduction from others, which are entirely different from it.[143]

Now, as has been pointed out, Hume's original articulation of the problem does not state that one cannot reason from is to ought, but rather that "'tis necessary that it shou'd be observ'd and explain'd." Nevertheless, his comment that it "seems altogether inconceivable, how this new relation can be a deduction from others, which are entirely different from it . . ." does seem to suggest that the linking is difficult, if not impossible.

Midgley accepts the claim that you cannot simply work from facts to values, but that something else needs to take place. As she has argued, the facts of our nature are such that we are in conflict with ourselves, and only reasoning as to what is more essential or important to our humanity will clarify what we ought to do. Facts are not enough. That, however, is far from the strong interpretation of Hume, in which facts are immaterial, the lifeless material on which our values are expressed. Midgley finds this kind of division absurd. We worry about whether a horse is unhappy when left alone, for example, because we know that a horse is a naturally social animal, needing to be with others. We do not decide to worry about the animal's happiness as an independent choice;

it logically follows from knowing that sociability is central to the animals' happiness.[144]

Although Midgley understands that facts and values are not the same thing, that does not mean that facts do not influence values, or that values do not influence facts. There is a relationship between them. Midgley suggests that it would have been far more fruitful for moral philosophy to have examined the interrelationship between the two than to build a barrier between them.[145]

From the facts side, Midgley states:

> But the facts of evolution cannot guide us directly. They matter only insofar as they can help us to understand our nature, our emotional and rational constitution. Yet our understanding of that *does* give us practical guidance. Facts about it are directly relevant to values. Values register needs. It is a mistake to suppose that there is some logical barrier, convicting such thinking of a "naturalistic fallacy." We are not, and do not need to be, disembodied intellects. We are creatures of a definite species on this planet, and this shapes our values.[146]

The divorce of facts and values is in fact a divorce of body from mind, as if they were two separate universes. But Darwin's revolution is a revolution that shows that we don't have a morality disconnected from who we are. Our morality is rooted in our needs and wants, in others' needs and wants, and in the interrelationship between them. There is a natural history to our wants. The idea of loving our children is not some abstract construction taken out of thin air, or through abstract reasoning. It is part of our moral structure because it is part of what it means to be human. That doesn't mean that we don't make choices, but those choices are rooted in a context of our humanity. As has been argued, different biology leads to a different morality. The facts of human nature matter to ethics:

> ... my point is the common-sense notion that our structure of instincts, as a whole, indicates the good and the bad for us. I am saying that, if Socrates is right in his facts, he is right in his argument. That is, if it is *true* that people are naturally inquiring animals, and if that inquiring tendency has a fairly central place in their natural structure of preference, then it follows that inquiry is an important good for them, that they ought not to stop each other from doing it (unless they have to), and should do it themselves, to an extent in proportion to the other things they also need to do. And so on for other tendencies.[147]

Notice how here Midgley's argument works from facts to values. If people are naturally inquiring animals, and that natural inquiry is central to who they are, it follows that inquiry is an important good—that is, it has value, and is to be valued. This is not all the work, since we still have not asked its value in relationship to other values, which is why moral philosophy is indispensable. She does make the case, however, that the facts of human nature are central to defining what is of value in human life, and in the human community.[148] The fact that a horse is by nature social means that, for Midgley, social is by definition a good for the horse. That good still needs to be integrated with other goods of the horse into a larger context of what constitutes the horse's good, but it does not need to wait for value to be externally placed upon it.

So why the historic divorce between facts and values? G. E. Moore, in his philosophical work at the turn of the twentieth century, memorialized the current radical split between facts and values, largely in reaction to social Darwinism's crude attempts to work from fact to value and to make reactionary claims as empirical fact. In response, Midgley argues, Moore tried to cloister individual values from the tyranny of society. By divorcing "facts" from "values," Moore was safeguarding the individual, making values an individual and private pursuit.

This move to make "value" something personal, safe from the claims of the larger society, in Midgley's evaluation, was connected to a second trend: the cloistering of "fact" into a neutral, supposedly scientific definition. A universal morality claims that there is such a thing as right and wrong; that some moral judgments are in a very basic way facts just like scientific facts—that it is objectively wrong to murder, for example. But as science, and increasingly physical science, became the only legitimate truth that we have, any other claim to truth was seen as subjective. Science, according to Midgley, retreated into a supposedly objective, neutral, factual world and values retreated into a subjective, individualistic one.[149] As neutral objectivity became the standard by which scholarship was deemed scientific, the social sciences and humanities also retreated into a cloistered neutrality. Moral philosophers insisted "that it was none of their professional business either to make moral judgements themselves or to help other people to make them"; the historian F. W. Maitland stated that "he never dealt in opinions, only in materials for the forming of opinions"; and, according to Midgley, the academic study of literature also separated itself into a supposed neutrality of description and analysis.[150]

Midgley claims, of course, that the neutrality of facts is as large a fiction as the complete subjectivity and relativism of values. Facts are not raw data independent of a conceptual scheme which shapes them.[151] We see the world through the lenses of our worldview. Truly neutralizing our value judgments is impossible. We need to be more aware of our conceptual schemes, to critique them, and when necessary to change them:

> The choice is not between integrating facts into one's world-picture and keeping them detached from it. It is between good and bad world-pictures. The impersonality required is not total detachment, because this is impossible. It is responsible objectivity—the far more difficult task of becoming more aware of one's world-picture, doing all one can to correct its more obvious faults, and showing it as plainly as possible to one's readers in order that they may know fully what they are accepting. As far as emotional tone goes, this calls for what Darwin offered: the very careful avoidance of all cheap and simple ranting which might carry people away to accept one's views wholesale, but also the full, scrupulous expression of attitudes and feelings which seem to one, after thought, to be called for by the subject-matter.[152]

Midgley therefore, is critical of evolutionary ethics not because they mixed facts with values, but rather because they got the facts wrong, and then subsequently reasoned poorly from facts to values.[153] They got the facts wrong because they adhered to a problematic world picture that biased their facts toward a problematic view of the world. Attacking that world picture, showing that it lies behind theories of "survival of the fittest" in first-generation Darwinism, and "the selfish gene" in second-generation Darwinism, is a central focus of Midgley's work:

> What bias, then, is now misleading us? I am suggesting that it is an unbridled, exaggerated individualism, taken for granted as much by the left as by the right—an unrealistic acceptance of competitiveness as central to human nature. People not only *are* selfish and greedy, they hold psychological and philosophical theories which tell them they *ought* to be selfish and greedy. And the defects of those theories have not been fully noticed.[154]

The separation of facts from values, of "is" and "ought," by cloistering facts from the values which give them meaning, leads to an ever-greater specialization less and less connected to the larger picture. Rather than keeping the world picture implicit, and in danger of becoming invisible, Midgley is arguing that the world picture needs to be made explicit, open and transparent for examination and critique. It is the world picture which informs our pursuit of knowledge; without it, knowledge becomes a random list of unassembled data.

Midgley maintains that this random list of data is exactly what has happened as knowledge has separated from meaning. Whereas knowledge was once understood to be connected with meaning, today knowledge has withdrawn into an endless compilation of overspecialized information. She argues that the result of this kind of knowledge is perhaps a larger collection of information, but a loss of the greater context which makes such information have meaning, and changes it from mere information into understanding. "Understanding," Midgley writes, "is relating; it is fitting things into a context."[155]

There is an important educational idea here. Critiquing Dawkins's call for poetry to take itself seriously and be based on scientific knowledge, Midgley returns the favor, by arguing that better science will take place once the larger field of life, in all its subjectivity, is understood as the necessary background in which science takes place.[156] For science to proceed, scientists need to be integrated individuals with a well-thought-out worldview. Children remember facts when they have a context within which to place them—when the facts have meaning as part of a larger story.[157] Midgley's claim that the structuring of a rich worldview is central to human pursuits is later presented as having critical implications for educational philosophy.

Midgley is not arguing that values precede facts, and certainly not that facts precede values, but rather that the two are intricately connected with one another. They are interdependent. Our facts are dependent on knowing what we are looking for, which only comes from having a general outlook. Our values are dependent upon the facts that we have, and these facts can change what it is that we value and why. Darwin was a great scientist because he had a worldview, and that worldview was not impregnable to his empirical studies of the natural world. His worldview allowed him to have a context for his research, but his research also changed his worldview. Darwin came to his work not as a professional scientist, isolating his private self from his professional self, but rather as a rich human being with commitments and concerns, whose scientific work was an expression of his larger humanity, not a neutral activity separated from it. Midgley is arguing that our work is connected to the pursuit of life's purposes, to our lives as a whole.

A TRANSCENDENT LIFE

Midgley's teleological view of the meaning of a human life rests on her view of innate human nature. Because we have a nature, our purpose is connected to being more fully human, fulfilling our natures. Because there are conflicts between different motives, however, meaning is constructed first through the cultural memory into which an individual is born, and then through the individual life and the conflicts which s/he resolves. The standard by which these conflicts are negotiated are the emerging whole of the person. The radical eclipse of some parts of our humanity at the expense of others, or an obsession with pursuing one motivation at the expense of all others, can be disastrous for the building of a fully developed, fully human being, and is the source of evil.

Such an explanation, however, is not the whole picture. It might be inferred from the above that the individual is the standard by which value is to be judged. Each individual has his/her own identity, based on his/her shared and unique innate nature, developed by individual choice, and independent of any larger context of consideration. As has already been alluded to, one can in fact find the roots of radical individualism within many of the ideas which are expressed by Midgley as well: that freedom is about setting each of us free from the constraints of corporate society and convention, and allowing us to pursue our own uniqueness, prescribed to us by nature and constructed by us in the value choices that we make among conflicting claims. Midgley, of course, although she is sympathetic to some of the themes, radically disagrees with such ultimately atomistic conclusions.

For Midgley, to be human is to be whole. But her view of wholeness, like Kropotkin's and Dewey's in the first generation, is not, and cannot be, one of being whole only within oneself. While she has a strong concept of individuality, her view of individuality cannot be divorced from the larger social and natural context of human life. For Midgley, to be human is to be a social being. Our humanity gains expression only in the context of community and social interaction. Our sociability, however, is part of a larger context of being a natural being, meaning that our humanity gains expression only in the context of the natural world of which we are a part. These two points converge into what I would describe as a strong religious sympathy in Midgley's work, both of which share in extending the human being outward into the world.

It is a biological fact that human beings are a social species. That did not necessarily have to be, although it has been shown that sociability and intelligence seem to be strongly linked and potentially coevolutionary.

Because they are a social species, it is a basic need of human beings to interact, socialize and build communities. Contrary to Rousseau's view, or alternatively Hobbes's, human beings were never isolated, atomized individuals who then chose to come together in communities for instrumental reasons. Sociability is part of our nature. It is part of the unique link between parent and child, where childcare extends far longer than with any other species, and is central to building human identity—both of the parent and of the child. It is part of our sexuality, where our sexuality is intimately connected with our sociability. It is part of our learning—of our language and play and curiosity which all lead us out into the world. Isolate an individual from other human beings, deny him/her acts of love and friendship, and one has been denied his/her humanity. We are born to be part of society. The atomized human being is a fictitious idea. It cannot be an ideal, and there is no reason why it should be.[158]

The Enlightenment desire to cut us off from the authority of the past had a historical context. There was a need to cut away from the overly oppressive authority of corporate society and to reflect upon age-old customs and truisms. Life was, in fact, stifling. The shift to the atomistic model was, according to Midgley, part of the attempt to protect the individual from the tyranny of traditional hierarchies in social relations.[159] Furthermore, this fear of the community, or of the state, continues to be a real one. As Midgley argues, freedom as a negative ideal, echoing Berlin, has a real role to play. States are oppressive. Convention is stifling. Marketing is manipulative. We need to safeguard the individual from the manipulations and false moralities sold to us by states or commerce.[160]

There is a limit, however. The price of freeing ourselves from the oppressions of the past has been the cutting of all ties, and indeed glorifying the isolated individual as the human ideal:

> It [freedom] no longer means freedom *from* or *to do* anything particular. It has spread itself to cover the isolation of the individual from all connection with others, therefore from most of what gives life meaning: tradition, influence, affection, personal and local ties, natural roots and sympathies, Hume's "sentiment of humanity."[161]

In the shift from the hierarchical structure of corporate society to the disembodied freedom of modern life, alienation and loneliness have emerged. Since human beings are, at their core, a social species, Midgley understands the loss of social ties as dehumanizing—a communitarian position with a Darwinian twist. Citing DeToqueville, she shows that the American ideal of individualism, rooted in the atomistic metaphors of the Enlightenment, already contained then the danger of social isolation of one from the other, and from the community.[162]

Wholeness, therefore, is not achieved as an individual isolated from the world, but rather as part of it. Society does not stand as some foreign creature apart from the individual and his/her freedom, but rather is a central part of what is necessary for a healthy human life. We extend out into the social world. It is the arena in which our humanity emerges.

That arena is not only the social world. The human being is at home within the natural world, and is part of it as well. S/he was born of it, and adapted to live in it. Our emotional structure is completely tied in with the sounds, smells, sights and textures of the world in which we live. Our dignity as human beings does not come in opposition to the world we live in, but as part of it.[163] In the same way that human beings need other human beings in order to be fully human, they need the natural world around them as well:

> This "whole person" of whom we have been talking is not, then, a solitary, self-sufficient unit. It belongs essentially within a larger whole, indeed within an interlocking pattern formed by a great range of such wholes. These wider systems are not an alien interference with its identity. They are its home, its native climate, the soil from which it grows, the atmosphere which it needs in order to breathe. Their unimaginable richness is what makes up the meaning of our lives.[164]

Midgley has strongly identified herself both with the animal rights/liberation movement and with the environmental movement, allegiances which are self-evident from her philosophy. The environmental crisis, for example, is a result of problematic philosophical assumptions deep within our culture, which see the world as alien matter which has nothing to do with human physical and spiritual life. Our choice of mechanized, atomized metaphors create the underpinnings for a worldview which has ignored the natural world as our home, ultimately imperiling life as we know it on the planet.[165]

Human beings are not isolated, closed off in a world to themselves. The result of such a view will inevitably be a loss of meaning in life since it has cut off the context within which meaning exists. When we sever our ties to the world, we sever our ties to the context which gives meaning to life. We are, in fact, not isolated, but part of a larger whole. We were created, or evolved, as part of that larger whole. We are not the center. When we begin to think of ourselves as the center, and the rest of the world is eclipsed, we lose sight of our place. Commenting on existentialism, Midgley writes: "The impression of *desertion* or *abandonment* which Existentialists have is due, I am sure, not to the removal of God, but to this contemptuous dismissal of almost the whole biosphere—plants, animals, and children. Life shrinks to a few urban rooms; no wonder it becomes absurd."[166]

The world is not dead stuff, but rather a living story of which we are a part. It is impossible to understand our part of the story without understanding the story as a whole. We occupy one corner of the story. We share the larger story line with other parts, which, like us, are also simultaneously ends in themselves. Our curiosity about the world is linked simultaneously to understanding each part on its own as an ends, but also as part of the larger puzzle which helps us to understand ourselves, as well.

Midgley's philosophy, like Kropotkin's and Dewey's in the first generation, has religious sympathies, and for reasons similar to those of Kropotkin and Dewey. Their view of human beings as social beings leads them out into the social world, and ultimately leads beyond it, beyond anthropocentrism into the world as a whole. Quoting William James, she writes that

> An attitude could usefully be called religious so long as it was one directed to the world as a whole, "about which there is something solemn, serious and tender." It must also be an attitude of acceptance, not rejection, and an acceptance which is not grudging but enthusiastic.[167]

Although she shies away from speaking of God, she is very critical of secular humanism which isolates the human being from the rest of the world:

> We need the vast world, and it must be a world that does not need us; a world constantly capable of surprising us, a world we did not program, since only such a world is the proper object of wonder. Any kind of Humanism which deprives us of this, which insists on treating the universe as a mere projection screen for showing off human capacities, cripples and curtails humanity. "Humanists" often do this, because where there is wonder they think they smell religion, and they move hastily in to crush that unclean thing. But things much more unclean than traditional religion will follow the death of wonder.[168]

Perhaps the idea of wonder best captures the outward-looking view which Midgley has of human life, and will be seen as central to educational philosophy. Wonder is not simply curiosity. Curiosity is wonder without awe and reverence. It has lost the wider context. The object of our curiosity is in danger of becoming something without value, our relationship to it that of having knowledge devoid of wisdom. For Midgley, there is a paradox in the relationship with others around us—people, animals, plants, mountains, and rivers, as examples. On the one hand, we experience wonder as we ponder something which is separate from us, something fundamentally different from us, with an evolutionary story and purpose of its own. And yet, simultaneously, we recognize that its meaning comes from the same story that human meaning comes from, and that our life's purpose is intimately connected to the same source. Children, poets and scientists—that is, human beings who relate to life with a sense of humility and awe—have a particular prescience for wonder.[169] Midgley claims that all three personality types are present in all of us.[170] One can assume that Midgley believes that all three were present in Darwin, as well. His wonder at the world—delighting in its otherness while simultaneously merging with its oneness—is in fact a fair definition of mystical experience. As Darwin says at the end of *On the Origin of Species,* "there is grandeur in this view of life."[171] And as Midgley states at the end of *Evolution as a Religion,* "As such, it is the business of each not to forget his transitory and dependent position, the rich gifts which he has received, and the tiny part he plays in a vast, irreplaceable and fragile whole."[172]

CHAPTER FIVE

A Darwinian Education

What does a contemporary educational philosophy rooted in a view of human life at home in the world look like? I argue that three central organizing ideas emerge from the Darwinism of Kropotkin, Dewey and particularly Midgley. The first is that Darwinism allows us to rebuild a framework for speaking of educational meaning and ends. This is especially interesting, since Darwinism is considered by many to have been a significant factor in the demise of educational ends, as I discuss herein. In the current debates around what is important to learn and why—that is, what are the goals of education—Darwinism offers a compelling vision of what constitutes a life worth living, and what education's contribution to such a vision is to be. The second organizing idea of Darwinism is that emotions and reasons are not separate universes, but are intimately interrelated. We are embodied beings, and therefore our reason is rooted in our emotional lives, and education needs to nurture their interdependency. Emotions are a central component of reason, guiding its work, and reason helps us to extend and structure our emotions. The third central idea is that we are social creatures, by nature. We relate to the world and gain meaning from within our social situations, but our natures nevertheless are the starting point for our socialization, and can resist socialization and reach out to the larger circle of life. In the debate between communitarian and liberal educators, Darwinism offers a third way, combining particular and universal identities, seeing us embedded in particular social contexts but having a common nature which allows us to reach out across the cultural barrier to other human beings (and other species). After expanding on each of these central ideas, I look at the implications of such an educational philosophy for

THE AIMS AND PURPOSES OF EDUCATION:
A DARWINIAN PERSPECTIVE

It is largely assumed that the contemporary world suffers from a loss of educational aims and purposes. Neil Postman, for example, wrote extensively about the loss of meaning in education, and argued that one of the chief slayers of traditional educational aims was Darwinism, which undermined the belief in a purposeful world, and offered little as an alternative.[1] Organizing narratives around which educational practice can be planned gives, as Postman states, "purpose and clarity to learning," and without them, education is without direction.[2] The educational philosopher Alven Neiman has similarly argued, quoting Dewey, that Darwinism did much to undermine the traditional idea of aims in education.[3] Although religious worldviews continue to hold onto teleological frameworks of meaning, the secularized world has lost these frameworks.

In *After Virtue,* Alistair MacIntyre gives a historical accounting of the demise of meaning. He describes the fall of the Aristotelian system, and with it the loss of the Aristotelian biological description of human needs on which the *telos* of a human life was based.[4] This contributed to the divorce between facts and values, so that biology, which was no longer rooted in Aristotelian science and thus could no longer describe human needs, was also no longer seen as having implications for the meaning of a human life. Values had once been considered facts—"this is, in fact, the essence of a human life"—and therefore the definition of the good life was supplied by the empirical facts of the species.[5] The fall of the Aristotelian system, in MacIntyre's view, led to a loss of moral direction as purpose was removed from biology, and our current moral intuitions are remaining fragments of a system whose justifying principles have been disproved.[6] Facts, divorced from values, became incoherent pieces of information, a point which Midgley makes, as well. Aristotelian philosophy was dependent on Aristotelian biology.

MacIntyre, however, does not abandon the Aristotelian project, but rather tries to resurrect it. He has tried to philosophically reconstruct teleological ideas of meaning, while accepting the collapse of Aristotelian biological essences. His intellectual project has been reconstructing an Aristotelian philosophy on the basis of a social rather than a biological view of human nature.[7] For MacIntyre, articulating the central critique communitarian philosophers have of liberal philosophy, humans are not atomized individuals who are open to choose freely their path in life. Instead, we are born into situations, and these situations form our life's starting points, its essences. We are born into a family, a culture, a place and a historical situation, to name some of the more prominent parts of our identities. Given that the essence of a human being is that of a social being, it is the familial, cultural and historical situation to which we are born and with which we are continually confronted that suggests and directs life's purpose and ends. MacIntyre attempts to decouple Aristotelian philosophy from biology, substituting social essences for biological ones.[8]

The Darwinism of Kropotkin, Dewey and Midgley is related to the communitarian positions, here and elsewhere, as we shall see. However, whereas MacIntyre and other communitarians reintroduce teleology without biology, Darwinism reintroduces biology to Aristotelian philosophy, using Darwin's theories of evolution as its new foundation. Meaning emerges out of our embedded natures. Because we are biologically a social species, life's meaning is to be found as a social species. For Kropotkin, as an example, the evolutionary strategy of mutual aid is the defining feature of human life. Since we are, at our essence, interconnected, life strategies which deny this biological reality are ultimately dehumanizing since they deny us what is essential to a human life. The atomized, competitive and selfish models of human life and cultures are both destructive to our humanity and ultimately unsustainable because they are contrary to our natures, and we will eventually rebel.

As described, Dewey also attempted to reconstruct a philosophy of meaning based on Darwin's biology. In his essay "The Influence of Darwinism on Philosophy," Dewey argues that Darwinism eliminated the teleological Aristotelian worldview where essences prescribed the proper path which a human life should take.[9] In a world of essences, the path of a life was prescribed by the essence of the species, actualizing its innate potential. According to Dewey, evolution teaches us that there are no essences, that the natural world is in a constant state of change and that it is impossible therefore to speak of any essence to a species, or any trajectory of a life dictated by actualizing its teleological ends.[10] Dewey believed that pragmatism offered a Darwinian alternative to Aristotelian teleology, substituting proximate ends based on the meaning in the moment, rather than ultimate ends based on unchanging essences. Dewey's project was the reconstruction of ends and purposes to human life, based on proximate Darwinian ends rather than ultimate Aristotelian ones.

Dewey's position is in some ways quite similar to MacIntyre's. Having defined the human being as, at its essence, a social being, Dewey goes quite far in the shift from natural essences to social ones. However, Dewey, unlike MacIntyre, never allows the social to exist independently of the natural. The social is a natural phenomenon, and therefore has its own biology. Habits, for example, have an evolutionary history. And the natural gives human beings their initial motives and desires, from which meaning is constructed. What we desire and want is originally given to us by birth; culture helps us to shape them, but does not create them. And, no less importantly, our innate natures restrain culture and afford us the resources with which to rebel against it when our socialization wanders too far from our innate humanity.

Whereas Dewey emphasizes the social, while maintaining the tension between the social and the natural, Midgley most forcefully emphasizes the natural side of the equation and its implications for a teleological worldview. MacIntyre's work offers a good point of comparison when exploring Midgley's philosophy. MacIntyre describes the Aristotelian system as consisting of three components: "untutored human nature, man-as-he-could-be-if-he-realised-his-telos, and the moral precepts which enable him to pass from one state to the other."[11] These are applicable to Midgley's Aristotelian philosophy as well. In Midgley's terminology, "untutored human nature" is prerational, innate human nature.[12] She holds onto, far more than Dewey, a rich and robust notion of an innate human nature as the evolutionary product of a social species. These innate motives form the goods of a human life. For Midgley, however, these are what I have described as goods in a weak sense. They are weak because they have not yet been structured into an integrated human life, and therefore remain as potential. In Midgley's language, "man-as-he-could-be-if-he-realized-his-telos" is a life lived as part of a larger whole. As a social species, our life's meaning is found in the context of the social and natural world. This point is critical to Midgley's philosophy, and as a result, to her educational philosophy as well. She claims that fulfilling our teleological ends as human beings is dependent on our understanding the human good as fundamentally connected to the good of others, and the larger whole.

I believe that this is the central educational aim of Darwinian philosophy—identifying and pursuing the teleological ends of a human life, which resides in the human connection with the world. As previously described, there is no doubt that in Midgley's philosophy the central goods of a human life are, as for Kropotkin and Dewey, intimately connected with our interrelationship with the world outside of the individual. When she writes that "we need this vast world," she means that our fulfillment as human beings is dependent upon connecting to the world around us. We need the world outside—"outer nature"—so that we can realize our inner nature. In Midgley's view, the idea that we could somehow live a fulfilled human life, developing to our human potential while being cut off from the world and its experiences, is preposterous. The nature within, human nature, is dependent on the nature without for its well-being, happiness and fulfillment. That nature without, it should be added, is not metaphoric. The loss of community, for example, is the loss of a critical circle where human beings actualize their humanity. We are not atomized individuals, but rather a social species who need one another to become who we are, and need a way of life in which we can pursue our basic goods for play, conversation, compassion, love and all the rest of the social species' repertoire.

The loss of the natural world is another example of the destruction of the context in which human beings become truly human. The dream often presented in science fiction literature, of moving civilization to another planet after humans have made this one uninhabitable, is absurd for many reasons, not the least of which is the absurdity of believing that a fulfilling human life can be lived someplace other than where human life evolved.[13] Even if we could figure out how to physically survive, the depravation experienced by removing human beings from their ecological home would be psychologically and spiritually devastating. One can describe this as an extreme form of alienation, where one is estranged from one's own humanity, since one cannot be human in an environment which is itself radically inhospitable to human life and living. The actual destruction of the natural world, therefore, is destroying our very ability to be human, by turning it slowly but inexorably into a foreign and alienating environment. Depriving human beings of contact with air, water, flowers, trees, animals and certainly other human beings is depriving them of the context in which to become fully human.[14]

But how does one deduce these ends from the weak goods of innate nature? Why should altruism and nurturing be seen as somehow more central to the ends of a human life than, for example, aggression or jealousy? In Midgley's system, as in Dewey's, motives are evaluated according to their role in the larger integration. Motives of selfishness, for example, might be part of the weak goods of innate human nature, but they are not a central goal of human life's teleology, because they undermine the quality of the whole. They undermine the ability of the individual to fulfill the central goods of a human life as a social species reliant on his/her relationship and identification with the larger whole.[15]

Dewey's reworking of Huxley's metaphor of the gardener is helpful here. The gardener does not accept innate nature as is, nor does s/he see it as simply raw material to be used. Rather, nature itself is used as both the material but also the guide for what the garden is to be. The gardener works with nature's innate potential and refuses to allow one part of nature to undermine the potential of the whole. Weeds in the Deweyian metaphor are comparable to human selfishness—a good in the weak sense but a bad in the strong sense, because they undermine the ability of the garden as a whole to flourish, and in the end undermine even its ability to grow. Innate nature directs the gardener, but the garden at its completion is not identical with innate nature left to its own accord.

As with Aristotle, Midgley's system rests on the ability to deduce from the weak goods offered by biological nature what exactly the strong goods are which make up the aims of a human life and therefore the aims of education. Following Aristotle, Midgley believes that the weak goods of life not only offer the raw material from which a life is built, but they also prescribe what those strong goods are which should be nurtured and developed through a lifetime. She, like Dewey, maintains that the weak goods offered through our biology clearly suggest what their correct integration into a strong set of goods should look like. The strong good of a life, for Midgley, is the integration of the weak goods in the multitude of ways which can connect us with others and the world around, and she believes that such a view emerges from a correct interpretation of what our biology itself suggests to us.

As Dewey suggests and Midgley makes explicit, just as it means something for a fox to act as a fox, or a gorilla to act as a gorilla, it means something for a human being to act as a human being, and conversely, for certain behavior to be inhuman, that is, foreign to how a member of such a species should behave. In a weak sense, any behavior of an individual of a species is self-evidently natural to the species. However, a fox acting like a fox or a gorilla acting like a gorilla means that there are behaviors which are unnatural, outside of what is understood as the correct integration of the multiple traits of a species into a functioning whole. While the fox and the gorilla might potentially have less of a gap between their innate natures and their integration into a functioning life, with different tools at their disposal for navigating the building of a life, they too must learn to become something slightly different than what would develop from their innate natures if left on their own.

I add here that Midgley supports the notion that there are biological differences between the sexes within a species as well, such that the traits and motives of men and women, boys and girls, are different. This should suggest educational implications, and is obviously part of the controversy surrounding Darwinian philosophy, and with it, its educational implications. There are, of course, differences between the individuals of a species, as well. Each gorilla, for example, has a particular personality which is the result of its innate characteristics, unique from within the gene pool, as well as its own particular experiences, also unique. The spectrum is widest among human beings, owing to their open instincts and tied to their ability to learn and adapt to new situations. The spectrum, however, is neither endless nor random.

This position, that there is a particular way of living that is measurably more human than others and, by extension, cultures that are more human than others, is obviously a highly contentious one.[16] The claim that there is a human nature on which our values are based was the heart and soul of social Darwinist claims about race, women, the poor and non-Western cultures. Blacks are born inferior to whites, women are by nature irrational, the poor are the evolutionarily unfit and aboriginal cultures are primitive. A Darwinian teleology seems to return us to such an essentialism, where dubious scientific claims justify societal discrimination. Is homosexuality, for example, unnatural, and therefore an abomination?

Criticisms of the teleological position come in three varieties. The first variety of criticisms holds that there is no essential set of characteristics which describe the human being, and so there is no description that can lead to a prescription for a human life. Human beings come in all shapes and sizes. Once again, using the example of homosexuality, humans have a large gene pool which consists of both homosexual and heterosexual behaviors. There are a multitude of "natural" behaviors, and therefore no human behavior is more natural than another. The second variety of criticism rejects the notion that values proceed from facts, so that any facts about human nature should have no bearing on what we value. We choose our values, and whether we are or aren't born with a sexual preference should have no moral implications for the sexual choices that we make in life. Whether or not homosexuality is "natural" should have no bearing on whether it should be viewed as moral behavior. Finally, the third variety of criticism rejects the idea that there is such a thing as values-neutral facts. Any movement from facts to values is flawed since the movement from facts to values is really a move from values couched as facts. Describing human beings as, by nature, heterosexual, is an ideological agenda, not an empirical fact. Social Darwinism, for example, moved from a values-driven set of supposedly objective facts to justify its worldview.

Midgley correctly accepts the first criticism, and assumes that humans have a wide spectrum of behaviors as a species. However, as argued, she rejects the assumption that there is an infinite variety of behaviors, and that it is therefore impossible to talk of a spectrum of behaviors which can be described as human. The challenge, to be discussed momentarily, is how to correctly describe the spectrum of behaviors which make up our species. As to the second criticism, claiming that facts can stand independently from values is a fiction. It is, in fact, impossible to actually keep values independent from the facts of our existence. All of our judgments, including our view of right and wrong, are shaped by the human lenses with which we judge. Our mind does not stand outside of those lenses, but is rather the quintessential organ of our humanity. It is a human mind. Darwin was the first to make this critical Darwinian point in the context of morality. As described in the introductory chapter, Darwin suggests that since human nature is different from other species, its sense of right and wrong is going to be a product of its particular nature, and would be different if its nature were something else. If bees were an intelligent species, he suggests that their morality would be based on bee nature, rather than human nature, and its sensibilities of right and wrong would emerge differently as a result. Midgley reiterates Darwin's point. If our thought cannot disconnect itself from our natures, it is logical that our natures are going to influence our moralities. Our morality, like everything else about us, is embodied. There is no neutral position from which we judge, but rather our views of right and wrong, good and evil, come from within our natures. Thomas Eagleton put it this way—"we think as we do because of the sorts of bodies we have...."[17] We value love because we are a creature that loves. I will soon argue that because we are capable of loving other human beings we also *should* value love. I am not making a philosophical argument here, however, but a biological one. We value loving because we are a creature that loves. If we were a different type of biological being we would be valuing a different set of attributes. Our biological natures already offer the categories within which we evaluate our vision of a good human being and a worthwhile life, and therefore the "is" must help determine the "ought." We cannot step outside of our natures.

Given that our needs as a species are results of our bodies, should we nevertheless attempt somehow to ignore our natures, attempting to separate the "is" from the "ought"? On one level, as just argued, that is impossible. Human beings are embodied beings, and therefore by definition everything that human beings do is "natural." However, while it is true that human beings cannot completely step out of their natures, the flexibility of our instincts combined with our consciousness and intelligence certainly allows a wide range of choice. It is possible to distance ourselves from our biological inheritance, even if we cannot free ourselves completely. The question then becomes whether we should choose to see our natures as an ally or an enemy. Culture shapes nature, like the sculptor who crafts a natural piece of wood, and so it is legitimate to ask whether we should relate to our natures as prescriptive in the process of sculpting a life, even if we are never actually independent from the natural world; therefore, choosing to ignore our natures is never really a complete option.

Why shouldn't we see our biological natures as prescriptive? Other things being equal, why shouldn't we accept as a reasonable and rather robust starting point that who we want to be is connected fundamentally to who we are? Why create a semi-fiction of our lives being somehow disconnected from our beings, putty in the hands of our culture, to do with what we want without direction from who we are? I find Midgley to be eloquent on this point. To ignore our biology and believe that we have total choice in deciding who we want to be is impossible. To attempt to reject it is radically misanthropic, reflecting a deep alienation from our very beings. It seems far more reasonable to see our social selves, the selves that we are in the process of becoming, as an extension of our natural beings, and not as a disconnected or alienated being in war with itself. Dewey argues exactly that when he states: "[Moral] rules can be obeyed and ideals realized only as they appeal to something in human nature and awaken in it an active response. Moral principles that exalt themselves by degrading human nature are in effect committing suicide."[18] Separating values from facts, therefore, is to resist what makes us human.

Finally, to the third criticism of teleological thinking, Dewey and Midgley (and in this they differ fundamentally from Kropotkin), accept the idea that facts are influenced by values, just as values are a product of facts. The "is" emerges from the "ought" no less than the "ought" emerges from the "is." It is not being overly postmodern to accept the commonplace that our ideological worldview shapes the way we describe the world. Breaking away from Darwinian positivists, Dewey states it clearly when he explains that we only have access to the world through human eyes. His pragmatist perspective suggests that all facts are value-laden and cannot be seen as objective and neutral. Since the world is a social world, which gains meaning for human beings through human experience, each object has meaning based on its effect on human experience. The candle, Dewey's paradigmatic object, is defined as giving light, because that is how humans experience it. The candle is also a source of pain; thus it gains a new, more integrated meaning based on a new experience. Every object, therefore, is pregnant with human meaning, the only meaning that we have access to, and its meaning emerges only in social interaction with other human beings.

Science, therefore, is not presenting some neutral, value-free truth about the world as it really is. That, of course, makes arguing from "is" to "ought" extremely dangerous, as it is using the supposed value-free authority of facts to push an agenda which is already implied within the facts. The Social Darwinists, for example, were not arguing from a neutral "is" to "ought." They were arguing from a set of ideological assumptions couched as scientific facts. Historians of science are continually uncovering the ideological assumptions which underlie scientific research and discovery.

Midgley suggests what I believe to be a perfectly reasonable middle position between positivist and deconstructionist hermeneutics. Midgley argues that we should accept the dynamic relationship between facts and values, constantly evaluating our worldview (values) in terms of its factual basis, and evaluating our facts by understanding the worldview on which they are based. Complete disclosure is called for, so that both our facts and our worldview are exposed and criticized, all the while knowing that they are from within, and not outside of, our subjective viewpoint. Democratic culture is therefore a central component of any healthy relationship between fact and value, since open and critical dialogue is a necessary component in the search for a healthy relationship between facts and values.

The difficulty with Midgley's position is, of course, that so much is at stake in our scientific description of human nature. Since it defines our good, it matters morally whether men and women are different from one another, and in what ways. It matters whether homosexuality is a biological trait or a cultural one. It matters whether men and women are different from one another, and how. Acknowledging that there are no objective facts that are accessible independent of human observation serves to underline just how complicated it is to implement Midgley's philosophy. Midgley's point, which I think is correct, is that we are ultimately safer in uncovering and acknowledging the ideological underpinnings of our science, and therefore always on the lookout for what motives underpin our scientific claims, while realizing that rejecting an innate human nature is just another scientific claim with ideological underpinnings, and not an alternative. Midgley, I believe, is quite forceful in articulating the price of the ideological claim that there is no human nature.

The linchpin in Midgley's teleological philosophy, and I believe the core of its educational implications, is in the movement from prerational human nature to the fulfillment of the human telos. For many animals there is only a small gap between their given nature and its fulfillment, and therefore only a minimal role for education. As Midgley points out, their instincts can be considered closed, meaning that they are preprogrammed for specific behaviors, with the environment often being nothing more than a trigger for innate behaviors to manifest themselves.[19] Particularly in the social species, however, as Dewey also points out, instincts become increasingly open, allowing for different strategies to develop in different situations.[20] Human beings, as can be witnessed in anthropological comparisons between different cultures, have such a wide range of behaviors that it seems hard to describe any particular behavior as more "human" than another one. Nevertheless, Midgley is claiming that while there are a large variety of behaviors, there are at the same time signature human qualities central to the species which need to develop in certain integrative ways for the human being to become fully human. The gap between our given natures, with their weak goods, and a fully formed human being who actualizes his/her human potential, the strong good, can only be bridged through education. Whereas closed instincts would negotiate the gap between "untutored human nature" and "man-as-he-could-be-if-he-realized-his-telos," open instincts decouple the two, creating the gap which makes education both possible and necessary.

I have shown how prerational human nature is seen as a guide for what human nature is then to become, negotiating the gap between biological human nature and the telos of human beings. But exactly how does that movement take place? The attempt to move from what human beings are to what human beings should be is where the educational act resides. Whereas the "man-as-he-could-be-if-he-realized-his-telos" describes the aims of a human life, the movement toward fulfilling those aims is the task of education. Like in Aristotle's system, here too there is a gap between the human being as s/he is and the human being as s/he should be, which is bridged through education, in the broad sense of adapting a human life to its purpose. How education is to move the human being from a loose integration of wants which is his/her biological inheritance, to a tightly integrated life interconnected with the world around him/her, can best be understood through understanding the interplay of emotions and reason, which I shall now describe.

EMOTIONS AND REASON

Aristotle, Dewey and Midgley all see a gap between innate human nature and actualizing the potential of that nature, and for all three that gap defines the role of education. Whereas the human *telos*—whether ultimate, as in the case of Aristotle, proximate and ever-changing, as in the case of Dewey, or proximate in evolutionary history but ultimate in human history, as in the case of Midgley—provides the goals of education, the gap between the potential and its development is what makes the educational process necessary. For Midgley, I believe that process is most clearly understood through her understanding of the interrelationship between reason and emotion, which together navigate the journey from innate human nature to a fully lived human life.

Midgley's use of the term *prerational* to describe innate nature might suggest that emotions are the starting point of a human life, but have little to do with navigating the movement from human nature to a human life, from "is" to "ought," which is the task of rationality. It is reason which shapes our emotions from their innate prerational phase to their mature rational phase, and therefore the development of our innate natures into a full human life is primarily a rational affair. Nature supplies the materials, rationality molds the outcome. This is far from the case, however, and a comparison with what Paul Hirst calls "the rationalist model" in educational philosophy helps to illuminate the embodied perspective of Midgley and other "embodied" Darwinists.

Paul Hirst, the British educational philosopher, outlines three periods in educational philosophy over the last fifty years, each with its own definition of educational goals.[21] The first, "the rationalist model" dominating the 1960s, defined educational aims as the development of rationality and the acquisition of knowledge. This approach saw the ultimate goal of education as developing human cognitive capabilities and knowing facts about the world. The development of human rationality, or reason, was therefore the aim of a good life, rationality being defined as the signature characteristic of a human life. All other human qualities were then judged as means to the ends in the development of coherent human reason.[22] For example, Scheffler could speak of cognitive emotions, showing how emotions can serve reason, and how, therefore, they do not need to be seen as reason's antithesis, necessary liabilities in the pursuit of reason, but rather as important tools in the achievement of human ends.

Although Scheffler rhetorically rejects the dichotomy between reason and emotion, arguing that reason is the ultimate of human emotions, it is clear that reason is something separate from the rest of the emotions, the arbiter of which emotions should be nurtured and which ones should be suppressed.[23] Peters suggests educating the emotions, once again based on the standard of reason.[24] For him, as for Scheffler, the emotions are a fact of human life to be shaped and utilized by the needs of cognitive reason.[25] Scheffler and Peters are examples of the rationalist model. In Hirst's view, the rationalist model gives a transcendental justification for the choice of rationality as the goal of human life, that is, an a priori privileging of reason as the central good in human life. The worth of other innate human tendencies is then evaluated according to the transcendental standard of rationality as life's ends.

Emotions, therefore, in the rationalist worldview, were often seen as value-neutral, needing to be evaluated according to their contribution to reason, or shaped in order to contribute to the cultivation of reason. The neurologist Antonio Damasio, in discussing the dominance of the rationalist model in neurobiology, argues that rationalists view instincts as disturbing the functioning of reason and decision making, which demand "a cool head," free of any emotional disturbance. Reason and emotions were viewed as two separate neural systems. Reason was clearly of higher importance in the hierarchy, and it ultimately served to tame and train the emotions.[26] In the rationalist model's influence on developmental theory, emotion was linked to a lack of maturity. The ability for rational thought emerged as the individual matured and was educated, and emotion was left safely compartmentalized behind.[27] Indeed, as the child grew, education placed less and less emphasis on emotional development.[28] The moral development of children was dependent on the development of their rationality, and the submersion of their emotional instincts, or at least their subservience to rationally defined aims.[29]

A Darwinian educational philosophy, I contend, stands in opposition to the rationalist educational tradition as outlined by Hirst. Midgley's view of emotions is central to her educational philosophy, and critical to Darwinism, as we saw in the discussion of first-generation Darwinist philosophy. Unlike the rationalist tradition, where emotions are either the neutral or negative side of human nature, which can be either enlisted or marginalized, emotions are at the heart of what makes us human, and what gives meaning to being human. Reason disembodied, that is reason without the emotions, transcendental in that it is seen to appear independently of an evolutionary history, loses the context which gives us what to think about and what to live for. Critiquing the rationalist tradition which sees human purpose in the cultivating of a pure intellect, Midgley whimsically comments that the dream of a mind freed from the body would leave us nothing to talk about, except perhaps mathematics.[30] If one needs to decide on a hierarchy between the emotions and reason, Midgley's philosophy sees human emotions and not reason as defining the human essence. Here she departs from Aristotle, who focuses on contemplation and its link to reason as what forms the ultimate *telos* of human beings, given that contemplation is unique to the human species, and uniqueness gives it primacy as the ultimate ends.[31] For Midgley, there is no contemplation without the emotions which shape and direct our reasoning. Those emotions, many of which are found in other social species, are central goods for humans, regardless of whether they are unique. It is an argument which we have seen made by Darwin, relying on Hume, and continuing through Kropotkin and Dewey.

Emotions are what give direction to human actions. Without them, human life loses its motivation and its compass. Whereas the rationalist model sees the direction of a human life through an emotion-less or emotion-served reason, Midgley contends that our emotions give structure and meaning to our actions. A life without emotions is one that lacks a meaningful structure from which to apply reason. In such a situation reason becomes lost at best, dangerous at worst. Midgley's definition of wickedness, and Darwin's and Dewey's too, is based on the absence of emotions, not their presence.[32]

It is the emotions which guide the individual from the weak goods of innate nature to their integration into a whole human life. Since the trajectory of a human life is not predetermined by closed instincts, human motives needed to be unusually strong in order to encourage human beings to act properly so that they could survive and flourish. Without them, open motives would allow flexibility in adapting to changing environments, a tremendous evolutionary advantage, but would not guide the individual to needed behaviors. Without strong motives of love and a need for companionship, for example, friendship would not take place, since a species with open instincts could choose to reject friendship as a good. Without a strong motive for nurturing, parenting could also be abandoned. As Midgley argues, it is because we have such strong motives that abandonment of such behaviors is the exception rather than the rule. Our innate natures offer us a strong guide from which to pursue life's goods. Our emotions, therefore, are central to the educational process of clarifying and pursuing the ends of a human life.

The moral life is the life that is lived in pursuit of the good life, that is, the life that a human being is meant to live. Education is about helping the individual identify the good and moral life, and offering the tools to pursue it. The good life is not discovered outside of the emotional life, as the rationalist model would suggest, but rather through its cultivation. Emotions, therefore, are both ends and means. They are ends in that they make up the goods of a human life. They are also means, however, in that they offer the individual motivation and direction in pursuit of their fulfillment. We are drawn to love, for example, and that allows us to obtain it. A rich, developed emotional life, therefore, is a necessary condition for a human life, and with it a moral life. Morality does not exist as something imposed on our innate natures, as the rationalist model suggests, but rather as part of our innate characters as a species. Applying Midgley's thought to educational philosophy, education has ignored the fundamental underpinnings of morality by associating morality with rationality and ignoring its emotive structure. The fact that emotional education is so consciously absent from school curriculums, for example, particularly as one advances in age, is a dangerous mistake of schooling, when looking at education through Darwinian eyes. The fact that male socialization in Western society encourages repressing emotions can be understood, therefore, as a dangerous mistake as well. Her view on emotions is far from the Freudian position, foreshadowed by Huxley and endorsed by Dawkins. Nurturing a wide variety of emotions, rather than suppressing them, would be Midgley's educational stance. This too, mirrors Dewey's position.

The proper development of emotional life is a necessary condition for pursuing the proper aims of human life, but it is not sufficient. It is clear that rationality plays a central role in Midgley's understanding of the movement from prerational nature to the fulfillment of the human *telos*. Rationality is needed in order to gain a clear understanding of what the purpose of a human life is, to be discerned from the empirical starting point of prerational nature. While the strong emotions of a species with open instincts can point us in the general direction, rationality is nevertheless needed to clarify the destination. Rationality is also needed on the journey, in order to consciously negotiate the conflicting internal claims on a human life's direction, and to prioritize them according to the goals of a life as they have been understood. Other species, with smaller gaps between their "is" and "ought," take the journey without the tools of rationality. Evolution added consciousness and rationality to the human toolkit, tools which are essential for the human pursuit of his/her ends. Rationality is a critical tool that humans use to close the gap between prerational human nature and how that nature needs to develop if it is to become truly human. Rationality completes the job which human emotions begin.[33]

Although Midgley attempts to do away with the dichotomy between reason and emotions, at times her description of them seems to fall into the dichotomy which she claims to be criticizing. Hirst's typology is once again helpful here. His second model, what he calls the utilitarian model, is presented as the antithesis to the rationalist model. In the utilitarian model, the educational emphasis is shifted from reason to emotion:

> In this picture, reason, knowledge, and understanding are not at all seen as themselves capable of determining from a detached point of view the ends that constitute the good life for individuals or society. They have the instrumental functions of helping us to discern, develop, and order coherently those basically given elements of wants and satisfactions from which the good life is to be composed. Nor does such a view see the dynamic of the good life as the exercise of the will enthralled in the service of reason. The dynamic is quite elsewhere, in the power over us of wants and satisfactions.[34]

Midgley's philosophy at times resembles Hirst's utilitarian model. Our wants and satisfactions (or, using Midgley's terminology—needs, goods, or desires) are the material from which the good life is composed. Reason plays an instrumental role, as it discerns, develops and orders coherently "those basically given elements of wants and satisfactions from which the good life is to be composed."[35] The emotions offer the ends to a human life; our reason offers the instrumental means by which to order our emotions into a coherent, functioning whole. Midgley at times describes the role of reason in exactly such instrumental terms.

According to Hirst, such a distribution of tasks between reason and the emotions maintains the dichotomy between the two, but reverses their prioritization. Emotions become the ends, and rationality becomes the means by which emotions are prioritized.[36] Reason continues to be something separate from the emotions, evaluating from above rather than from within. Reason is disconnected and disembodied from the emotional processes which give meaning, as it objectively stands above and evaluates.[37] Midgley at times seems to suggest exactly this approach, for example when she writes: "Intelligence alone would not generate these ends [of a human life]. It would just calculate means."[38] In such a view, intelligence is not an ends in itself, as Aristotle maintained, but a tool with which to arbitrate a hierarchy of ends generated elsewhere.

Although there are certainly instances where Midgley describes reason in the ways which Hirst critiques, I believe that Midgley ultimately must be seen as rejecting a view of rationality as a disembodied tool which stands outside of the body's emotions, as a neutral, transcendent observer.[39] In spite of her occasional descriptions of reason as a neutral tool for ordering the emotions, her philosophy as a whole suggests viewing reason as embodied as well. As she points out on numerous occasions, there is no neutral observer who can objectively evaluate. Our subjective lenses shape our observations, and therefore there can be no separating of our rationality from our values, which, as Midgley continually points out, have their source in human nature with its particular set of emotions serving as the goods of human life. Although at moments Midgley slips into a description of reason disconnected from our emotions, with mind separated from body, it is quite clear that her philosophy as a whole claims otherwise. It would make more sense to interpret Midgley as essentially breaking down the dichotomy between emotions and rationality, and to accept Midgley's explicit claims for reason emerging from our emotions. As she argues, thought completes feeling, which suggests that it is its continuation and not its outside evaluator. There is no isolated reason apart from the context of the body and our emotions.[40] There is no ghost in the machine.[41]

Reason does not only emerge from our emotions, but also shapes them. Both Aristotle and Dewey develop this point as central to their educational philosophies through the concept of habits.[42] Darwin also develops this point and argues that instincts are not only innate, but can also be learned. Gould described this process as biological potentiality. Although it is less developed in Midgley's philosophy, it does appear, and is obviously necessary to her system. Habits are consciously chosen behaviors which, having been performed repeatedly, eventually become a prerational instinct, the same as biological ones. There is no need to think about them; one eventually acts as if they were innate. Both Aristotle's and Dewey's education of character focus on building correct habits, which, while contrary to certain innate desires of the individual at first, are eventually adopted as an inherent good. Similarly, Midgley sees reason as developing habits which later become prerational nature, therefore bridging innate nature and human ends. Reason reshapes our innate nature to lead us to pursue our ends as human beings, with our reason and emotions working toward a common goal. Emotional education, therefore, should not be seen as being opposed to rational education, but rather as an integrated view of reason made up of emotions, and emotions shaped through reason.

Ultimately, therefore, Midgley's view of education toward the good and moral life, following Aristotle's and consistent with Dewey's, sees right behavior as dependent on right motives and temperament. Morality is not built only on right actions which are independent or contrary to our motivations, as certain rational models might claim. Morality is not just a question of outward behavior toward others. Rather, our emotional makeup and our reasoning are linked, such that our motives point us toward the right behavior and support that behavior. There are moral instincts, and they need to be taken seriously. They guide us toward the right behavior, that is, the behavior which is right for our species. Moral instincts are social instincts which stand at the core of human nature. Midgley's philosophy suggests that human beings ultimately need moral motives to do moral acts, and therefore education must tend to our moral instincts as well as to our moral actions. An Aristotelian view doesn't see our reason and emotions in conflict, but rather their seamless working together, where our morality is an extension of our emotions, not their adversary.

Yet, while emotions offer us a guide, morality is not reducible to directly acting on our emotions, given that our emotional needs conflict with one another, and rationality is a tool which evolution gave us for sorting between goods, and developing the strong good. Our emotions offer us guides for what is proper behavior, but they can also mislead, particularly if led to develop in discord with the individual as a whole, connected to his/her larger context. The task of reason, guided by our emotions, is to discern what the right behavior is, what the virtuous individual is, and to encourage motives which allow such a character to grow from the human potential, through the formation of habits which train our innate nature to lead us in the right direction.

Habit formation, however, is not only a cerebral process. Rationality helps us to identify what the aims of a human life should be and allows us to sort between conflicting motives and to choose those that more centrally support the overall aims of human life. Training of the emotions, however, demands more than the rational knowledge of how to act. As Midgley points out, evil is often conducted by those who know right from wrong very well, but whose characters are too weak to act properly. Aristotle's pedagogic suggestion, which Midgley supports, is that one should practice doing right acts, and their practice shall awaken our innate, but perhaps latent or retarded, desires to pursue such acts.[43] Our reason can train our actions; knowing the good can lead us to do good. Doing good can lead us to want to do more good, as it stirs that set of motives already there, which however have not played a dominant role in our personalities. Having done good acts, we shall discover that we enjoy doing such acts.

Ultimately Midgley, like Aristotle, Darwin, Kropotkin and Dewey, sees the moral individual and the happy individual as one. Tension between emotions and right actions is a temporary situation that education seeks to reduce. Reason puts our emotions on the right path, helps us merge our often conflicting goods into a coherent, functioning whole, and as a result, to become solid, happy, moral people. Reason takes the goods of our emotional lives and helps to weave them into a happy and responsible life within the larger web of life, the life that as social, conscious beings we were born to live.

An embodied Darwinian educational philosophy, the movement from "untutored human nature" to "man-as-he-could-be-if-he-realized-his-*telos*," can be summarized as consisting of a number of interrelated steps:

1. The development of a robust emotional/rational understanding of the weak goods which make up the individual's life.
2. Gaining an understanding of how these weak goods should be formed into a strong individual identity, based on our interrelationship with the world around us, in order to become truly human. Our emotions need to be trusted as a reliable, but not faultless guide.
3. The discerning of the right behavior as a function of the balancing of competing goods in the context of our emerging identity as a whole.
4. The acting on our reasoned conclusion.
5. The enabling of our acts to shape our emotions by becoming habits, so that those aspects of our emotions which support such right actions will be stirred to indeed support our at first perhaps passionless or even emotionally resisted deeds.

Such a description of educational philosophy, however, seems to suggest that the individual acts independently of society, as if the process were internal to the individual. This is, of course, not consistent with an embodied Darwinian philosophy, in which the individual is in need of the larger context in order to become fully human, or where habits are passed on within the culture, and adopted prerationally as second nature to the individual. And yet, if the individual only becomes fully human within the context of society, the individual stands in danger of losing his/her ability to resist culture, and to change it. Kropotkin, Dewey and Midgley were all aware of the conservative nature of acculturation, the dangers of particularist loyalties, and they all sought ways of maintaining a healthy dialectic between them. It is an issue which is intensely debated in educational philosophy today, and to which Darwinism has much to contribute.

PARTICULARISM AND UNIVERSALISM

Culture is an essential component of Midgley's educational philosophy, as it is for Kropotkin and Dewey. In Aristotelian terms, culture is the historical transmission of one group's attempt to define the human *telos*, and it offers tools, such as ritual, with which the individual benefits from the collective experience.[44] The multifaceted manner in which cultures differ shows the many avenues within which the *telos* can be pursued, and the plurality of values supported by culture shows the many ways in which the *telos* has been understood in a given culture. Habit formation is not primarily the work of any particular individual, but rather the work of culture over generations. While individuals are born with an innate biological nature, they are also born into a culture, and they inherit characteristics, as well.

Acculturation, therefore, plays a critical role in education, in that it directs human nature to a certain set of culturally conceived answers to the human dilemma of conflicting motives. The collective wisdom of culture in general, or Western culture in particular, is not an obsolete and unscientific body of irrelevant traditions, as Keynes claims the Bloomsbury group mistakenly saw it.[45] For Midgley, it is rather the historically compiled knowledge of a group of human beings' struggling to build a vision of the good life, and a way of achieving it. From Dewey's perspective, it is the collective's record of problem solving. We are born into traditions, and it would be horribly arrogant to claim that they are unnecessary and that each individual must start from scratch. Initiation into one's own culture is an absolutely central part of the educational process.

However, like Dewey and Kropotkin, it is clear that Midgley is aware of the socialization dilemma. On the one hand, socialization is absolutely essential in order that innate human nature can even begin to emerge, and yet, on the other hand, it is, as Dewey understood it, a conservative activity which resists change. For Dewey, the didactic of problem solving is what forces the individual to question cultural assumptions, and to change cultural habits in order to behave according to newly understood meaning. Habits allow us to shape our given natures, but since the habits then become prerational, they are conservative in nature. The realization that we are confronted with a problem, an ambiguous meaning, means that our habits can move from the prerational, unconscious and instinctual, to the rational, conscious and reflective. The realization of the problem wakes us up from our instinctual behavior and forces us to reevaluate cultural assumptions, therefore allowing for cultural change.

Dewey seems to suggest that human nature offers the engine for such a process, giving us a natural curiosity and motive to find meaning. Otherwise Dewey is confronted with the dilemma of where the individual gains a perspective other than his/her culture which allows him/her to understand a cultural assumption as problematic. Kropotkin also maintains a human nature apart from socialization, which allows the individual to resist socialization when it goes down the wrong path.

Midgley's solution to the acculturation dilemma is more robust. She has, as described, asserted that our given human nature offers a place from which we resist and can adjust the process of socialization. Our given human natures allow us a place from which another voice may be heard. It anchors a view of what it means to be a human being. Without it, the human being is defenseless against the assumptions of society. It is only with a view of what it means to be human that we can evaluate if cultural practices do damage to our humanity. She claims, for example, that the idea of dehumanization can only make sense if we have a concept of what it means to be human, which Midgley contends that we gain from the innate nature of our species.

The liberal/communitarian debate and its implications for educational philosophy can be helpful here. One of the central critiques by the liberal position of the communitarian perspective is about the danger of parochialism.[46] MacIntyre, for example, argued that humanity's moral compass and life's meaning is gained from the social and historical context in which s/he is born. The liberals then ask, "Where does one then get the ability to step out of one's culture and evaluate its truths?" In the liberal tradition, from which Hirst's rationalist model emerges, our reason allows us to reflect and transcend our situations, and to see the world and our culture from a critical position. The liberals contend that the communitarian position does not offer an alternative place from which to criticize and change culture.

For example, in her article on educational philosophy, "Patriotism and Cosmopolitanism," Martha Nussbaum sees cosmopolitan reason, freed from the particularity of culture, as enabling us to be both moral agents but also subjects—an ends with intrinsic worth.[47] She contests the claim of communitarians who see moral meaning embedded in our situations, and claims that only the outer circle of our humanity is a morally relevant category, owing to our shared capacity as rational beings.[48] She, therefore, advocates a universalism, with the primacy of reason as the defining characteristic of the moral life, rather than particularism with its communitarian commitment to, in her eyes, morally irrelevant categories of family, community and country.[49]

Midgley is also aware of the parochial potential of acculturation being the only source of our worldview. Having rejected the rationalist position, however, she solves the dilemma through her view of biology. A communitarianism grounded in a view of the human being as a social being indeed risks relegating the human being to his/her social situation, without a place from which to resist its norms and habits. But Midgley's view of human nature offers a place to resist the parochial, not through a disengaged reason that stands outside of the embodied situation, but through an innate human nature with its moral intuitions, part of our biological inheritance as a social species. Innate human nature stirs our critical senses, offering an inner voice which tugs us in a different way when our socialization is radically at odds with our natures, and can serve to offer reservoirs for resistance. The natural offers a place to guide, and if necessary to resist, the cultural.[50]

Of course, such a view is directly at odds with social constructivists who hold that meaning is always constructed through interaction with society, and therefore there is no such thing as precultural, or prerational values. Dewey's formulation might be more acceptable to constructivists—that there are motives, but these motives only gain a coherent shape through their interaction with culture. But, as I have pointed out, it is a mistake to identify Dewey with at least radical social constructivists. Like Kropotkin before him and Midgley after him, he too held that the innate motives contribute to raising questions. Impulses independent of socialization raise questions which suggest a problem that needs to be solved, and they also contribute to formulating solutions to new problems which have surfaced. The Darwinian perspective sits at the crossroads between the essentialist and constructivist position. Accepting humans as social beings, it recognizes that meaning is mediated and emerges from the social world. However, claiming that there is a strong human nature, inherited at birth, it maintains that socialization takes place in interaction with an innate nature which is always present and active. Midgley's hermeneutic, like Dewey's, can be described as a dialogical one, or, utilizing Gallagher's typology, a moderate hermeneutic.[51]

In effect, it is our moral emotions, and not our disengaged rationality, which allows us to jump from the inner circle of family to the outer circle of humanity. We are preprogrammed to reach out in sympathy and affection to others. How far that instinct takes us through the concentric circles of association is an important question. According to some interpretations of evolutionary theory, any group loyalty necessitates a rival group against which group loyalty enables a selective advantage.[52] Our innate sense of group loyalty, therefore, will always come at the expense of our connection with a larger circle of humanity. If we follow this line of reasoning, human beings have an innate proclivity to, on the one hand, identify with a particular group and, on the other hand, to demonize those outside of their group, who will be considered foreign and thus suspect. Midgley calls such a process pseudo-speciation—that is, taking the idea of a barrier between species, which is a real biological boundary, and applying it to differences between groups of human beings, where it is a social boundary, albeit emerging from biological motives.[53] The barrier between groups, though not a given at birth, is part of the socialization process to which humans, being a social species, are susceptible. As Midgley says, echoing the evolutionary research, "the more richly a social bond develops, the greater, inevitably, is the difference between those inside it and those without." [54]

The point for Midgley, however, is that our sociability, our need for others, is prior to socialization within any particular group loyalty. At first, the sense of belonging is to other human beings, and only later to a particular group.[55] Our innate natures draw us to other humans; our socialization into a group shapes our identification with others into a sense of belonging to a particular family and group. Its implication for Midgley is that our innate social natures, presocialized, allow us to reach out to the largest circle of humanity and beyond, but, as it is cultivated within a particular context, ultimately distances us from identification with the larger human community.

Because for Midgley group loyalty is a characteristic which is learned, it is part of the socialization process. She believes that young children, however, who are first being initiated into inherited characteristics, still hold onto their innate propensity to identify with other human beings, as in fact they do with other animals, particularly the other social species who return in kind.[56] They smile, interact and respond. This innate ability, however, is slowly eclipsed by socialization, as the child's innate identification with other beings is replaced by particular human beings. Nevertheless, they can, if they are able to maintain that childish predilection, continue to recognize their commonality with humanity, and indeed with other animals.[57] As Midgley would argue, the educational challenge is to allow our intuitions to exist somewhat independently of the necessary but also at times damaging process of acculturation. The child, still intuitively open to the larger world, maintains moral intuitions which can guide him/her into an adult life where the child within is not ignored or suppressed but is celebrated as an essential part of an adult life.[58] Whether such intuitions can be nurtured through education as a resource to resist socialization to dehumanizing attitudes and behaviors is a critical test for the application of an embodied Darwinist philosophy to education.

In conclusion, we can see in such a Darwinian educational philosophy, emerging from Midgley's philosophy, with roots in Kropotkin's and particularly Dewey's thought, three interlocking educational components. First, Darwinism provides a biological framework from which we can return to an Aristotelian discussion of the *telos*, where the proximate aims of Darwinism substitute for the ultimate aims of Aristotle's biological theory. Education is to help us close the gap between the raw potential of our innate natures and the integration of that only partially shaped nature into a coherent whole which, like Dewey's gardener, works from within the inner logic of the raw material to actualize its potential. The integrated ends of a human life are constructed from the ends which our biology gives us. Human ends, being part of a social species, are located in association and integration with the world which is necessary for human fulfillment of its ends, and gives meaning to those ends. Second, because our reason is not transcendent of our biology, but emerging from within it, our path to the human *telos* is guided by both our emotions and our reason, or, perhaps more accurately, by our reasoned emotions, which can be trusted but still are shaped along the way as we grow from our potential to its actualization. Education is not about repressing our evolutionary emotional heritage, or building a conception of the good and moral life without any help from our natures, but rather it is about using our emotional nature as a guide which is then shaped as its inner conflicts are resolved. Habits allow our reason to shape our innate emotions. Lastly, culture is an essential tool in human development toward its *telos*, since human nature is biologically dependent on social interactions to properly develop. Culture is, no less importantly, the predominant vehicle for understanding our *telos*, since it is the collective wisdom of a particular people's exploration of how a human life should be lived. No one individual can, or should, start the process from scratch. Nevertheless, there is the danger of acculturation preventing a critical voice from developing which enables individuals to see outside of their particular culture and allow culture to be criticized and to change. Kropotkin, Dewey and Midgley all contend that our innate human natures offer a moral instinct which allows us to resist culture when it moves to forms that are dehumanizing, that is, against our nature. Strongest in our childhood, before socialization has overwhelmed it, ideally it is fostered and developed by culture but also remains as a wellspring from which to oppose culture, if necessary.

Having summed up the central implications of an embodied Darwinian philosophy for educational philosophy, I now look at how it can be expressed in curriculum and in a didactic approach. I first examine its curricular implications, and then explore a central implication for teaching methods.

FROM NATURE TO CULTURE: A DARWINIAN CURRICULUM

I would suggest that a Darwinian curriculum can be divided into three interlocking components, paralleling my suggestion for how to view educational philosophy. The first component would be subjects that helps students clarify their *telos* and pursue them; the second would be subjects that foster the integration of our emotions with our reason, so that our emotions can play a central role in guiding the pursuit of our *telos*; and the third would be a curriculum which cultivates community while simultaneously allowing us to avoid the dangers of particularism, and which allows us to powerfully identify with the social and natural world of which we are a part. As in a Darwinian educational philosophy, each of these curricular pieces is interconnected, and they build on one another.

1. Poetry and Science: A Curriculum Which Clarifies the Telos

As Midgley argues, our innate natures can guide us toward the *telos*, but they cannot do all the work. If they could, one could assume that there would be no need for structured education. As Spencer argues in the first generation, education takes place all the time in nature without schooling. Birds learn to sing their mating songs, mice learn what and how to eat, and wolves learn to hunt.[59] As Midgley points out, however, the more open-ended these behaviors are, as they are in humans, the more they can go wrong.[60] Their strength, as Dewey emphasizes, is in their incredible flexibility, allowing human nature to adapt to a wide range of experiences, and therefore, fundamentally, to express itself in a wide range of behaviors, far wider than any other species. Their weakness, Midgley asserts, is that without a strong guide to motivate in the proper directions, human life can easily be derailed.[61] Education, it can be deduced, comes to strengthen the natural instincts in order to guide human life.

According to Midgley, culture can be seen as a particular historical group's answer to what constitutes the good life. The young, therefore, do not start the process from scratch. Education can first and foremost be seen as the initiation of the young into a cultural worldview, into literature and philosophy and music and history. Such an educational philosophy clearly suggests the critical need for a humanistic education. A humanistic education can offer what Midgley calls a road map, that is, a view of life's destination, and a route to get there. In Midgley's view, as in Dewey's, individuals eventually build their own road maps, but they rely heavily on the cartographic work of the previous generations.[62] Without such a map, which she alternatively calls a worldview or paradigm, knowledge loses its anchor and becomes atomized pieces of information rather than the wisdom which allows information to have meaning by being part of a larger puzzle.

Midgley's hermeneutic underpins this approach. Since she rejects the modernist fact–value distinction, where science is the arbiter of [objective] fact, and the humanities deal with [subjective] values, she cannot rely on a curriculum which gives prominence to the facts [science] as the only knowledge that really matters, or as science and the humanities as separate universes, each cloistered from the other, or even as facts defining the scope within which values are then applied. Rather, all information is dependent upon its context to evaluate its meaning. Without a larger worldview, there is no clear way of connecting the information, determining what is pertinent and what isn't, and scientifically recognizing what one is looking for. Midgley describes Darwin's efforts to construct his own worldview to illustrate her point:

> When the young Darwin immersed himself in the arguments about cosmic purpose in Paley's theological textbook *The Evidences of Christianity,* and repeatedly read *Paradise Lost* on exploring trips from the Beagle, he was neither wasting his time nor distorting his scientific project. He was seriously working his way through a range of life-positions which lay on the route to the one he could initially use.[63]

Darwin's worldview, of course, was not an objective fact of the world, but rather an organizing metaphor, capable of changing when challenged with discrepancy from the empirical information which justifies it. Being well read, attending to the larger picture, and examining competing versions of the larger picture were all necessary steps to Darwin's theory of evolution, according to Midgley. Science exists within a culture, not separate from it. Studying the worldview, therefore, and building one's own, is central to being able to navigate the path to a truly human life, which is the goal of education. One cannot do without a worldview; it is only a question of whether one critically attends to it or not. The humanities are central to this purpose.

Science serves several purposes. In the context of defining our *telos*, it is a central ingredient in making sense of our lives. Since in Midgley's system our values are interconnected with the facts of our lives, the description of those facts is of profound importance for understanding our lives and how we should lead them. Darwin's theory of evolution emerged within the context of an education which gave Darwin a background into which his scientific work could operate, but his theory also changed his and others' worldviews. There is a dialogue that takes place between the empirical description and the larger worldview when defining life's meaning. The worldview allows us to organize information, but once organized the information can challenge the larger view, and change it. Science is not a description of an absolute truth about the world, to which the humanities are seen as subjective and therefore peripheral. Science is, however, in constant dialogue with the larger worldview and affects it. The forest and the trees need each other. Once again, Dewey and Midgley's moderate hermeneutics suggests a dialogue between facts and values, where facts are mediated by values, and values emerge from facts.

The second and related role of the curriculum in the pursuit of the *telos* is in pursuing our human ends, and not only in defining them. Music is a necessary part of the curriculum as an ends in itself. It enriches a human life, because it is a need. (Remember, no natural susceptibility, no Beethoven.[64]) Art is also an innate need.[65] Schooling allows us to develop our natural desires for music and art by initiating us into a rich cultural heritage which has refined and expanded these most basic desires. The pursuit of theoretical mathematics might very well be acting on a basic human need to satisfy the intellect, but it is certainly not the only, or even the central, human ends. We have multiple ends, and education should encourage us to continue their cultural development as part of a larger view of what a human life is. Midgley, for example, is concerned when the pursuit of one particular good—say, art or math—eclipses all other pursuits and displaces the pursuit of a whole human life.[66] Her philosophy leads to a curriculum that should primarily allow us to build our view of the good life through cultivating the wide variety of motives with which we are born and which our cultures have extended.

Midgley is very clear about the usefulness of uselessness in the curriculum. The term "use," she argues, has two meanings which are relevant to education.[67] The first relates to "use" as an instrumental means to different ends. Education, for example, is a means to finding employment, and math, therefore, is a necessary part of the curriculum because its basic mastery is indispensable for many jobs. Learning trigonometry is only useful to those who pursue becoming an engineer; a foreign language for those who pursue diplomacy or international business. In this view, philosophy is largely a useless subject. "Use" can also refer to something of value, however, and value, as Midgley describes it, is connected to ends, not means, being parts of a larger whole.[68] Midgley's curriculum, therefore, puts a tremendous emphasis on exactly those subjects that in an instrumentally driven curriculum would have little place. It is exactly because they are useless—that is, an ends and not a means—that they are most valuable. As she attacks the instrumental notion of "use" for education, she argues that when education focuses solely on training for employment, without tending to human life and its manifold needs as ends, one will find despair, alienation, and depression, with their concomitant failure in the workplace.[69] An ends-driven "useless" education might also be the most useful of educations, nurturing meaning and motivation.

2. Developing Emotional Intelligence

Because our goods are rooted in our innate natures structured around our emotions, it is clear that emotions are not irrelevant to the curriculum, but rather stand at the heart of it. The rationalist model of education essentially ignored the emotions, since educational aims were defined according to dispassionate and neutral reason, acting independently of our emotional evolutionary heritage. The goods of a human life are expressed in our innate motives, which are at their core motives of a social species manifested in interaction with others.

What this means is that the development of our emotional lives, essential for the development of our moral instincts and the pursuit of the good life, is dependent on the development of our social relations. We become fully human in the context of others. Our humanity emerges as we reach out into the world and recognize that our needs are interwoven with the larger whole. We need friendship, affection and love. We have an innate motivation for sympathy and compassion for others. Aggression, jealousy, and competition are central components of our being. Education, far from developing only the mind, must be about educating the whole person, of which our emotions are a central ingredient. Our reason is as dependent upon our emotions to engage it as our body is dependent upon our emotions to mobilize it.

For such a social species, social interaction becomes critical to the curriculum. As Midgley argues:

> ... for a full life, a developing social creature needs to be surrounded by beings very similar to it in all sorts of apparently trivial ways, ways

which abstractly might not seem important, but which will furnish essential clues for the unfolding of its faculties.[70]

Without such rich interactions, our emotional and intellectual lives will be retarded. Midgley quotes John Stuart Mill, who in his autobiography suggests that English culture as a whole suffered from exactly such developmental retardation:

> I did not then know the way in which, among the ordinary English, the absence of interest in things of an unselfish kind . . . and the habit of not speaking to others, nor even much to themselves, about the things in which they do feel interest, causes both their feelings and their intellectual faculties to remain undeveloped . . . reducing them, considered as spiritual beings, to a kind of negative existence. . . . But I even then felt, though without stating it clearly to myself, the contrast between the frank sociability and amiability of French personal intercourse and the English mode of existence, in which everybody acts as if everybody else (with few or no exceptions) was either an enemy or a bore.[71]

As a test case for such an approach, it is helpful to explore a curricular subject such as physical education. As shown in the section on Herbert Spencer, Darwinian ideas had an influence on how Spencer perceived the role of physical education in the curriculum, based on his Darwinian materialist view of the body's health being central to the survival and thriving of the species. In Spencer's model, physical education was about promoting the physical fitness of the species in order to facilitate the evolutionary process. By contrast, in Midgley's model, physical education can be understood as developing game-playing and the interactions among a social species, as they explore, for example, the roles of competition and cooperation in their lives. It is social, not physical, education. Current education largely ignores the potency of such educational situations, as teachers remain ignorant and ill prepared to explore and enrich the emotional learning which is largely taking place as part of a "hidden curriculum" in physical education.

This approach associates Darwinian educational philosophy with educational philosophers such as Nel Noddings and Jane Roland Martin, who advocate education as a nurturing experience and domestication as an important framing metaphor for education.[72] Arguing against the rationalist model of subject matter, Roland Martin argues that:

> One finds repeated demands for proficiency in the three Rs, for clear, logical thinking, and for higher standards of achievement in science,

mathematics, history, literature, and the like. One searches in vain for discussions of love or calls for mastery of the three Cs of care, concern, and connection.[73]

Beck and Kosnick structure the emotionally rich class and school community into three clusters which need to be nurtured: a community of rich conversation; a community of celebration, joy and openness; and a community of tenderness, security, friendship and mutuality.[74] Furthermore, they argue that "emotional education" should not be defined as a separate subject, but rather should be woven into the very heart of the school's culture.[75] As shown, emotions are not a separate category from intellect and culture but rather the driving force in its motives and workings. Literature and history as subjects are about emotional education at the same time as they are about building worldviews and values. As Roland Martin argues, the curriculum of traditional cultures "encompassed myth, ritual and custom and initiated the young into ways of perceiving, feeling, thinking, acting behaving toward others and toward nature itself."[76] The development and pursuit of an integrated emotional life is pursued within culture, not apart from it.

3. Concentric Circles: A Curriculum That Reaches Out Into the World

Although culture is indispensable to the development of a human life, it is not alone responsible for generating life's meaning. Our innate human natures, while ultimately shaped by the surroundings on which they are dependent for proper development, nevertheless offer a starting point from which culture works. In the concentric circle model posited by some communitarians, culture shapes humanity through interactions and allegiances to a tight-knit group of people—family, friends, community—in which human sympathies develop.[77] There is, however, a danger with this. As Midgley points out, "the more richly a social bond develops, the greater, inevitably, is the difference between those inside it and those without."[78] While sympathy, therefore, begins in the innermost circle, it must be educated to move outward through the circles. Midgley, however, believes that precultural sympathies in the child begin with immediate connection to the largest circles, and are necessarily narrowed to a smaller circle only later. Adoption works, for example, because the infant will respond to anyone who is able to understand his/her needs well enough to fulfill them.[79] The curricular challenge is maintaining the more universal sympathy, while at the same time nurturing that sympathy in concrete cultural contexts.

Midgley believes a positive effort is needed to resist what she calls pseudo-speciation, the belief that there is a fundamental barrier separating one's closest circle of humanity from other circles:

> On the one hand, even in explicit terms of language, not all humans admit all others as belonging to their species at all. The arrangement whereby the name of one's own tribe is also the word for "human being" seems to be quite widespread. This can—though perhaps it need not—mean that outsiders are treated as "only animals," notably in the two important respects of hunting them down without hesitation and rejecting with horror the idea of intermarrying with them. This attitude is part of the larger phenomenon called "pseudo-speciation"—the tendency for human beings to regard their cultures as if they actually were separate species. Pseudo-speciation is what makes it possible for "cultural evolution" to proceed so fast. Customs of all kinds are accepted by imprinting early in life, taken for granted as a part of one's constitution, in a way which makes it easy to go forward and make whatever new inventions are needed to supplement them. The price paid for this, however, is that people with different customs tend, at a glance, to seem like members of a different species, and so to be rejected. Of course it is possible to resist this process, but a glance around the world makes it clear that this needs a positive effort. Baboons, presumably, do not ask themselves whether the baboons of a strange invading band are really proper baboons at all, but people can ask this question, and can answer it with a no.[80]

Multicultural education is often used as a tool for developing empathy beyond one's particular social group. Nussbaum, for example, believes that multicultural education is critical for human beings to discover the similarities, and not only the differences between human beings, and to build allegiances to a greater humanity.[81] The channeling of one's innate sympathies for other human beings to sympathy for only a particular set of human beings needs to be resisted. Our innate moral sympathies make that not only possible but also desirable.

The outer circle, however, to which our nature not only can but also should draw us, does not end with other cultures. It reaches out beyond the species barrier. Not only do our sympathies with other human beings need to be nurtured, but so do our sympathies for all other living things. Our desire and need for connectedness to others is primarily filled by our connection with other humans. The species barrier, unlike the division between other human cultures and races, is a real one, and exists among all species.[82] We are drawn to those whom our emotions recognize as connected to us, and who respond in kind. Midgley uses the hypothetical example of intelligent species from outer space who are incapable of supplying the needs of humans, because they are incapable of recognizing them and responding to them. We recognize human suffering because we understand human emotions as our own, and therefore recognize that pain and humiliation does indeed hurt.

Yet, because we are a social species evolved from the same evolutionary story as other social species, the species barrier is not nearly as impenetrable as it might be with a species foreign to our world. We are cousins to other species and therefore the barrier can be crossed constantly. Domestication of animals demands that on some level we know what it feels like to be a cat or a dog, and that they are capable of responding and building a relationship with us. It should not be surprising that domestication takes place in social species, not only because of their intelligence, but also because of their need for sociability which human companionship can partially fulfill. Children, as Midgley points out, cross the species barrier all the time, as kids and puppies and kittens recognize each other's natural predisposition for play. In the wild, animals have been known to adopt the young of other species.

A school based on an embodied Darwinian educational philosophy would be filled with animals for play and companionship. A Darwinian philosophy clearly recognizes, for example, why animal therapy works. But, just as a Darwinian educational philosophy suggests interaction with the animal world, it also suggests learning about the animal world, like the multicultural social world, as a way of learning about our own humanity. In the same way that Midgley relies on ethology to teach us about both what is unique among humans and what is common to us and others, so too a Darwinian curriculum would rely on ethology to teach us about ourselves through the study of other species. The study of primates, birds, and other social mammals all offer a window into knowing ourselves. Self-knowledge is dependent upon understanding our connectedness to the world around us and seeing ourselves through the prism of our place in the tapestry of life.

The concentric circles of the curriculum, therefore, work simultaneously at a number of different levels. We can and must build our individual identities in relation to each of the circles, but they do not have to extend outward linearly. Each circle can play a unique role in building our bonds while defining our own uniqueness. Animals are not only a mirror for understanding ourselves, nor are they relevant only insofar as they are similar to us. Their uniqueness as something different, and yet part of the same story, is what helps us to understand ourselves as also different and yet connected. The same is true for other cultures.

Ultimately, a Darwinian curriculum rests on ecology and evolution as critical subject matters, which teach us the interdependence of all living life, and that humans are but one part of a larger story. There is a reason why we think a rose is beautiful and its aroma is sensual. It is not a social construction. The source of delight resides as a central part of our innate nature.

CULTIVATING WONDER: EDUCATIONAL DIDACTICS

I have already surveyed some of the didactical implications of a Darwinian educational philosophy in the section on curriculum. Teaching, for example, needs to be aware of emotional education at all times, which demands emotional awareness, the ability to articulate and process the emotional richness of an educational community and, of course, knowing how to channel the emotions of the moment into an integrated life built on our emotional and physical interdependence. Kropotkin, as another example, examines the didactic implications of humans as social beings. Education must take place within a social context. Doctors need to be shown the social meaning of their work. Teaching scientific facts apart from their application will lead to a spiritless practice of medicine. Education needs to take place in the world, within the social contexts of our lives, and not in classrooms which, like the existentialist's lonely room, teach us that learning sits apart from life. The medium is the message.

Perhaps the heart of a Darwinian educational approach can be interpreted as based on the need to teach wonder, and its centrality in teaching an ecocentric view of the world. Midgley is careful to differentiate between curiosity and wonder. Curiosity, as she understands it, is a concept rooted in the myth of a detached, impersonal science. The ideal of science, for example, driven by the myth of the scientist's insatiable curiosity to know, abandons any context for the knowing and any purpose in its pursuit.[83] Science teaching, when not of the instrumental use-value variety of economic progress, is usually of this mode. Midgley instead argues for love as a bedfellow of curiosity, and suggests that together they form the proper motive of wonder. Curiosity without love is a dangerous proposition, as it is pursued independent of any other values which would give meaning and direction to its pursuit.[84]

Love is about relationship and connection, caring and responsibility, which stand at the heart of such an educational didactic. We need to build caring relationships with the world around us. Those relationships expand outward, beyond the species barrier and into the natural world. When left to look at human life in only its social context, we lose the larger framework which explains and gives meaning to the human story:

> The Darwinian perspective on evolution places us firmly in a wider kinship than Descartes or Hobbes ever dreamed of. We know that we belong on this earth. We are not machines or alien beings or disembodied spirits but primates—animals as naturally and incurably dependent on the earthly biosphere as each one of us is dependent on human society.[85]

Experiencing wonder is knowing that we are connected to something greater than ourselves, something which is both connected to us but also not reducible to us. Wonder expands the circle, moving us out of the human circle into the larger story. Wonder is the educational experience which moves us outside of our self-referential anthropocentrism, and lets us place life into its wonder-filled context:

> ... wonder involves love. It is an essential element in wonder that we recognize what we see as something we did not make, cannot fully understand, and acknowledge as containing something greater than ourselves. This is not only true if our subject-matter is the stars; it is notoriously just as true if it is rocks or nematode worms. Those whose pearl is the kingdom of heaven, or indeed the kingdom of nature, follow it because they want to drink in its glory. Knowledge here is not just power; it is a loving union, and what is loved cannot just be the information gained; it has to be the real thing which that information tells us about. Nor can the point be only the effect on outward action, though certainly it is true that if the love is sincere action will follow. The student will learn the laws and practise the customs belonging to the kingdom of heaven or of nature, trying to become more fit to serve it. But first comes the initial gazing, the vision which conveys the point of the whole. This vision is in no way just a means to practical involvement, but itself an essential aspect of the goal. On it the seeker's spirit feeds, and without it that spirit would starve. The desire for that vision is a desire to be fulfilled by it, which means accepting that, as one is, one is imperfect and needs to be made whole.[86]

Our inner nature requires the nature outside of us to complete it. In the rationalist model, an unbridgeable barrier separates humanity from the rest of the world, a barrier which inevitably leads to losing life's meaning by placing humans in the center of the story without the contextual backdrop. In Midgley's model, the individual can only develop his/her humanity by recognizing that it is a small, and in the spirit of Job, a rather insignificant corner of the larger story, a small part of a larger whole: "We need the vast world, and it must be a world that does not need us; a world constantly capable of surprising us, a world we did not program, since only such a world is the proper object of wonder."[87] Midgley sees Aristotle and Darwin endorsing Job's insight.[88] It is a religious sentiment passionately argued by Kropotkin and Dewey, as well.

How is wonder to be taught? One can imagine stargazing, observation of ant colonies and detailed descriptions of flowers as central parts of the curriculum. But as Midgley suggests in her analysis of the components of wonder, it is not the subject matter on its own which brings the message, but rather a particular attitude to life which must pervade the teaching. It is not only what we teach, but how we teach. Without love, for example, science is curiosity without values; with it, science becomes a "reverent understanding of the universe."[89] Science, from this worldview, becomes a central part of a religious education, an education which teaches that human beings are not at the center of the universe.[90] For Midgley, three types of individuals understand this truth—the scientist, the poet, and the child.[91] Education's role, paraphrasing Midgley, is to nurture the child to develop his natural sympathies into adulthood, so that the poet and scientist can live in each of us as mature expressions of that intuitive childhood experience of wonder.

FINAL THOUGHTS

As presented earlier, Neil Postman argues that contemporary education desperately needs a metaphysical framework, a framework which will give purpose and clear values to education, something in his view which is sorely lacking from our increasingly bureaucratic and instrumental view of education.[92] Can Darwinism do the work for educational philosophy that Postman claims education so desperately needs? Darwinism, as Postman himself points out, is popularly associated with the breaking down of worldviews, leaving in its wake a world without meaning.[93] I have tried to show that this is not the only interpretation open to Darwinism, nor is it the interpretation which is truest to Darwin's own interpretation of his work's implications. Kropotkin, Dewey and Midgley all claim that Darwinism can in fact serve as a paradigm which returns meaning to our worldview.

Darwinism goes a very long way. It offers a view of being human which gains meaning from being part of the evolutionary story. We are not blank slates to be molded, or independent beings whose identities are to be constructed. Rather, we are born with a biology which has a history, and which connects us viscerally to the only world in which we are created to live, and to which we therefore have an interest, as well as an obligation, to protect. Our biology manifests itself in our cultures, since, as social beings, we become human only in the companionship of others. Through culture, our biological identities are shaped and formed. This story gives direction to our lives and to our education, which is the cultural means for bridging the gap between our potential as human beings and the building of our identities. There is a *telos* to human life—a direction for each individual human life and for human beings as a whole, but the individual and culture are still left to navigate the gap between who we are potentially, by virtue of our biology and our culture, and who we are meant to be. Education is to help us steer the course, by assisting us in clarifying where we have to go through building and constantly challenging our worldview, and in how we are to get there, by integrating our motives into a coherent identity. By shaping an ideal of a human life, and enriching our ability to be more fully human, education creates a place where the individual and society can resist forms of dehumanization and alienation. Darwinism, far from the reactionary agenda which it is often perceived to uphold, is in fact an essential ingredient in the struggle to maintain human dignity in the face of dehumanizing cultural forms and directions.

Darwinism has been perceived as an anti-religious worldview. But if we define religiosity as understanding that we are part of a larger whole which gives us meaning, and the experience of that transcendence in our lives, then Darwinism surely advocates a religious worldview. Science does not stand in opposition to religion, nor independent of it, but as a central tool in teaching wonder, awe and reverence, and approaching the world with wonder is a necessary ingredient in true scientific pursuit.

Alan Ryan, in his biography of Dewey, challenges the notion that evolution can offer as much guidance as Dewey claimed:

> It is hard to recapture exactly the mood in which Dewey wrote about evolutionary ethics and the place of evolutionary considerations in philosophy generally. It is especially hard if one is as skeptical as I am about the capacity of the theory of evolution to do the philosophical work Dewey wanted it to do. Dewey and others were plainly right to think that many of our abilities and many of our desires are more or

less "built in" to us for survival-oriented reasons. What is less easy to see is how many desires and abilities and how much knowledge can be so accounted for.[94]

The same challenge to Dewey can be made to Midgley and other Darwinists. Can evolution as a worldview extend beyond the immediacy of basic survival functions and help us to explain our cultures and values—what they are and what they should be? Midgley has made a strong case that it can. Ryan and others are doubtful. In many ways the question is an empirical one, as science continues to use the Darwinian paradigm to explore the place of music and dance, language and communication, community and cooperation, love and affection, sympathy and altruism in human nature. In other ways, it is a question of values—does Darwinism tell a story which gives substance and meaning to a generous and rich human life? From this perspective, education could certainly do worse than nurturing a worldview, anchored in science but reaching beyond, which lets us embrace our humanity and have confidence in our species' innate nature to find life's meaning in our connections with those around us with whom we share life.

Notes

CHAPTER ONE. THE MAKING OF DARWINISM

1. The anthology by Amelie Oksenberg Rorty, exploring differing philosophers' and philosophies' implications for educational philosophy, applies a methodology similar to the one I use in exploring in Chapter Five what Darwinist philosophy suggests about educational philosophy. See Amelie Oksenberg Rorty, *Philosophers on Education: New Historical Perspectives, Philosophers on Education* (London and New York: Routledge, 1998). On Plato, see Zhang LoShan, "Plato's Council on Education," in Rorty, 32–50. On Aristotle, see C. D. C. Reeve, "Aristotelian Education," in Rorty, 51–65. Also, Samuel Scolnoicov, "Plato," in J. J. Chambliss, ed., *Philosophy of Education: An Encylopedia* (New York and London: Garland Publishing, Inc., 1996), 483–89; J. J. Chambliss, "Aristotle," in Chambliss, 30–34; Donald Abel, "Human Nature," in Chambliss, 282–85.

2. On Descartes, see Daniel Garber, "Descartes, Or the Cultivation of the Intellect," in Rorty, 124–138; and Peter Schouls, "Descartes," in Chambliss, 143–46. On Hobbes, see Jeremy Waldron, "Truth, Publicity and Civil Doctrine," in Rorty, 139–47; and Bernard Gert, "Hobbes' Psychology," in Tom Sorell, ed., *The Cambridge Companion to Hobbes* (Cambridge: Cambridge University Press, 1996), 157–74. On Locke, see John W. Yolton, "Locke: Education for Virtue," in Rorty, 173–89; and Peter Schouls, "Locke," in Chambliss, 362–66. On Hume, see Annette C. Baier, "Hume on Moral Sentiments, and the Difference They Make," in Rorty, 227–37; and David Fate Norton, "Hume, Human Nature and the Foundations of Morality," in David Norton, ed., *Cambridge Companion to Hume* (Cambridge: Cambridge University Press, 1993), 148–81. On Rousseau, see Amelie Oksenberg Rorty, "Rousseau's Educational Experiments," in Rorty, ed., 238–54; and David Owen, "Rousseau," in Chambliss, 566–73.

3. Stephen Jay Gould, "The Pleasures of Pluralism," *The New York Review of Books*, www.nybooks.com, June 26, 1997, 10.

4. Allen et al., "Letter," *The New York Review of Books*, November 13, 1975, 182, 184–86. On biological potentiality, see Stephen Jay Gould, *Ever Since Darwin:*

Reflections in Natural History (New York: W. W. Norton and Company, 1977), 251. For Gould's own articulation of the sentiment first presented by his colleagues in the 1975 letter to the editor of *The New York Review of Books* quoted above, see Gould, *An Urchin in the Storm* (New York: W. W. Norton and Company, 1987), 112–15.

5. Richard Dawkins, *The Selfish Gene* (Oxford: Oxford University Press, 1989 [1976]).

6. See Peter Singer, *A Darwinian Left: Politics, Evolution and Cooperation* (New Haven and London: Yale University Press, 2000).

7. Charles Darwin, *On the Origin of Species* (Oxford: Oxford University Press, 1996; reproduction of second edition, 1860), 394. Scholars, however, have argued that in *On the Origin of Species* Darwin does deal with the implications of his theory for human beings and society. It is argued that, in spite of the fact that Darwin wanted to bracket the question of human origins, it was so integral to the subject that it could not be avoided, and indeed already appears in *On the Origin of Species*, although it is not emphasized. See Kathy J. Cooke, "Darwin on Man in *On the Origin of Species*: An Addendum to the Bajema-Bowler Debate," *Journal of the History of Biology* 23, no. 3 (Fall 1990): 517–21.

8. See J. C. Greene, "Darwin as a Social Evolutionist," *Journal of the History of Biology* 10 (1977): 1–27.

9. Ernst Mayr, *One Long Argument: Charles Darwin and the Genesis of Modern Evolutionary Thought* (Cambridge, MA: Harvard University Press, 1991), 35–47.

10. For a short discussion as to what degree Darwin's theory was indebted to his reading of Malthus, see Mayr, 85–86.

11. Darwin, *On the Origin of Species*, 38–107; also Mayr, 35–47.

12. Darwin, *On the Origin of Species*, 53.

13. Darwin's adoption of Spencer's phrase 'survival of the fittest' was actually at the urging of Alfred Russell Wallace, with whom he shared the discovery of natural selection. Cited in Carl N. Degler, *In Search of Human Nature: The Decline and Revival of Darwinism in American Social Thought* (New York: Oxford University Press, 1991), 61. For Darwin's various descriptions of the meaning of the struggle for survival, see Darwin, *On the Origin of Species*, 53.

14. Ibid, 58.

15. Ibid, 53.

16. Ibid, 73–75.

17. Ibid, 73–74.

18. In his concluding sentence to the introduction to *On the Origin of Species*, Darwin states, "Furthermore, I am convinced that natural selection has been the main but not exclusive means of modification" (Ibid, 7). For his expression of frustration that this qualifier was largely ignored by his critics, see Darwin, *The Descent of Man*, v–vi.

19. Stephen Jay Gould, for example, seeks to emphasize that Darwin saw that there were other mechanisms at work in addition to natural selection, in order to dislodge character traits of animals and humans from a 'survival of the fittest' origin. In the introduction to all editions of *On the Origin of Species*, until the seventh, and last edition, Darwin states: "Furthermore, I am convinced that Natural Selection has been the main but not exclusive means of modification" (Darwin, *On the*

Origin of Species, 7). In the last edition (1872), Gould quotes Darwin, stating: "As my conclusions have lately been much misrepresented, and it has been stated that I attribute the modification of species exclusively to natural selection, I may be permitted to remark that in the first edition of this work, and subsequently, I placed in a most conspicuous position (namely at the close of the Introduction) the following words: 'I am convinced that natural selection has been the main but not the exclusive means of modification.' This has been of no avail. Great is the power of steady misrepresentation" (quoted in Stephen Jay Gould, "Darwinian Fundamentalism," *The New York Review of Books*, June 12, 1997).

20. Quoted in Mary Midgley, *Evolution as a Religion*, 6. According to Midgley, Lamarck also held this view, as did Spencer and many other of Darwin's contemporaries. She calls this view the Escalator Fallacy.

21. See Arthur O. Lovejoy, *The Great Chain of Being* (Cambridge, MA: Harvard University Press, 1964), 59–66. For the original Biblical sentiment, see Psalms 8:4–9:

> "When I behold Your heavens, the work of Your fingers,
> the moon and stars that You set in place,
> what is man that You have been mindful of him,
> mortal man that You have taken note of him,
> that You have made him little less than divine,
> and adorned him with glory and majesty;
> You have made him master over Your handiwork,
> laying the world at his feet,
> sheep and oxen, all of them, and wild beasts, too;
> the birds of the heavens, the fish of the sea, whatever travels the paths of
>   the seas."

22. Lovejoy, 315–33.
23. J. C. Greene, 1–27.
24. Darwin, *The Descent of Man*, 2.
25. Ibid, 65–98.
26. Ibid, 70.
27. Ibid, 80.
28. Ibid, 162. Darwin's use of the term 'mutual aid' in the preceding quote was later taken by Kropotkin as the title for his work on evolutionary theory.
29. See, for example, R. Axelrod and W. D. Hamilton, "The Evolution of Cooperation," *Science* 211 (1981): 1390–96.
30. Ibid, 166.
31. R. L. Trivers, "The Evolution of Reciprocal Altruism," *Quarterly Review of Biology* 46 (1971): 35–57.
32. See David Kohn, ed., *The Darwinian Heritage* (Princeton: Princeton University Press, 1985).
33. Charles Darwin, *The Expression of the Emotions in Man and Animals* (Chicago: University of Chicago Press, 1965, first published in 1872).
34. Darwin, *The Descent of Man*, 72.
35. Ibid, 73. Darwin's full quote is as follows: "I do not wish to maintain that any strictly social animal, if its intellectual faculties were to become as active and as

highly developed as in man, would acquire exactly the same moral sense as ours. If, for instance to take an extreme case, men were *reared* under precisely the same conditions as hive-bees, there can hardly be a doubt that our unmarried females would, like the worker-bees, think it a sacred duty." Notice that Darwin says that "if men were reared under precisely the same conditions" and not "if men were born with precisely the same nature" as the bees. Darwin often blurs the distinction between innate instincts and learned behavior, as is discussed later, so that human beings are, for Darwin, a consequence of both nature and nurture. E. O. Wilson and Mary Midgley emphasize the innate instinctual side of the quote in their work, writing explicitly of an intelligent species with a different nature having therefore a different morality. I believe this to be a fair interpretation of Darwin's intention, when seen in the larger context of his argument.

36. In general, one can see Darwin's ethics as an extension of Hume's perspective—the continuity with the animal world; the centrality of emotions, as opposed to disembodied reason; the primacy of sympathy within moral life; the determinist nature of a morality which emerges from one's innate nature, rather than from the supposed free will of rationality. See David Hume, *A Treatise of Human Nature* (Oxford: Clarendon, 1964 [1739]). Darwin himself cites the similarity between his argument and that of Hume. See Darwin, *The Descent of Man,* 110.

37. Ibid, 106. Darwin was aware that he was going against the philosophical mainstream, disapproving, for example, of both Kant's position (Ibid, 70) and that of J. S. Mill, whom he quotes: "'if, as is my own belief, the moral feelings are not innate, but acquired, they are not for that reason less natural.' It is with hesitation that I venture to differ from so profound a thinker, but it can hardly be disputed that the social feelings are instinctive or innate in the lower animals; and why should they not be so in man?" (Ibid, 71).

38. Here Darwin is explicitly attacking Adam Smith, who Darwin cites as arguing "that the basis of sympathy lies in our strong retentiveness of former states of pain or pleasure. We are thus impelled to relieve the sufferings of another, in order that our own painful feelings may be at the same time relieved." It is a hedonist argument, in that it is claiming that supposedly acts of selflessness are in reality self-centered acts of relieving our own pain, and not that of others (Ibid, 81).

39. Ibid, 91.

40. Ibid, 87, 91.

41. Ibid, 91. Or, again: "Thus at last man comes to feel, through acquired and perhaps inherited habit that it is best for him to obey his more persistent instincts. The imperious word ought seems merely to imply the consciousness of the existence of a persistent instinct, either innate or partly acquired, serving him as a guide . . . we hardly use the word ought in a metaphorical sense, when we say hounds ought to hunt, pointers to point . . . if they fail thus to act, they fail in their duty and act wrongly." This hints at a teleology that will reappear in second-generation Darwinism, in Mary Midgley's philosophy. "Hounds *ought to* hunt" because that is what they were born to do, even though other, conflicting instincts might misdirect them (Ibid, 92).

42. *Nicomachean Ethics, Book Two*: "the pleasure or pain that actions cause the agent may serve as an index of moral progress, since good conduct consists in a

proper attitude towards pleasure and pain"; in Aristotle, *Ethics* (New York: Penguin Books, 1982, from 1953 translation), 95–97.

43. On Darwin's general appreciation of Aristotle, including his moral philosophy, see Allan Gotthelf, "Darwin on Aristotle," *Journal of the History of Biology* 32 (1999): 3–30.

44. Darwin, *The Descent of Man*, 92.

45. This, too, is in keeping with Aristotelian views. See, for example, in *The Nicomachean Ethics, Book Two*: "our virtues are exercised in the same kinds of action as gave rise to them" (Aristotle, *Ethics*, 94–95).

46. Darwin, *The Descent of Man*, 89–90.

47. Ibid, 92.

48. Ibid, 92.

49. Ibid, 99.

50. Ibid, 81.

51. Ibid, 99.

52. Ibid, 164. Earlier, Darwin also states: "but it is worthy of a remark that a belief constantly inculcated during the early years of life, whilst the brain is impressible, appears to acquire almost the nature of an instinct; and the very essence of an instinct is that it is followed independently of reason" (Ibid, 100).

53. J. B. Lamarck, *Zoological Philosophy* (Chicago: University of Chicago Press, 1984), 122, as explained in Gillian Beer, "Introduction," Darwin, *On the Origin of Species*, xxi.

54. For a full exploration of the history of Mendel's scholarship, see Robin Marantz Henig, *The Monk in the Garden: The Lost and Found Genius of Gregor Mendel* (New York: Houghton-Mifflin, 2000).

55. Eitan Avital and Eva Jablonka, *Animal Traditions: Behavioural Inheritance in Evolution* (Cambridge: Cambridge University Press, 2000), 314–15.

56. In second-generation Darwinism, a neo-Lamarckian perspective is making a comeback, showing how culture in both human beings and other animal species allows Lamarckian evolution to take place, through the cultural passing down of learned behavior and habits from one generation to the other, and even the possibility of their being integrated into the individual as a biologically inherited trait. For an excellent presentation of neo-Lamarckian theory from within a Darwinian paradigm, see Avital and Jablonka, *Animal Traditions*, 304–51.

57. Ibid, 89.

58. Ibid, 104.

59. Darwin, *The Descent of Man*, 35. See also Kenan Malik, *Man, Beast and Zombie* (London: Weidenfeld and Micolson, 2000), 109–14.

60. Charles Darwin, *The Descent of Man*, 575–76.

61. See, for example, Sherry B. Ortner, "Is Female to Male as Nature Is to Culture?" in Michelle Rosaldo and Louise Lamphere, eds., *Woman, Culture, and Society* (Stanford: Stanford University Press, 1974), 67–87.

62. Darwin, *The Descent of Man*, 177.

63. As Darwin argues, "the careless, squalid, unaspiring Irishman multiplies like rabbits: the frugal, foreseeing, self-respecting, ambitious Scot, stern in his morality, spiritual in his faith, sagacious and disciplined in his intelligence, passes his best years in struggle and in celibacy" (Ibid, 174).

64. Ibid, 168. Elsewhere, Darwin points out approvingly that artificial selection is constantly taking place, for example when we incarcerate or execute individuals with 'bad' qualities, or when psychologically dangerous individuals are removed, or remove themselves, from society: "In regard to the moral qualities, some elimination of the worst dispositions is always in progress even in the most civilized nations. Malefactors are executed, or imprisoned for long periods, so that they cannot freely transmit their bad qualities. Melancholic and insane persons are confined, or commit suicide" (Ibid, 172).

65. Ibid, 177, 169.

66. Alfred Lord Tennyson, "In Memorium A. H. H.," in Norman Page, ed., *Tennyson: Selected Poetry* (London and New York: Routledge, 1995), 106.

67. Jean-Jacques Rousseau, *Emile, or On Education*, Introduction, Translation and Notes by Allan Bloom (New York: Basic Books, 1979).

68. For example, in Robert Ardrey, *The Territorial Imperative* (London: Collins, 1967), 288–305, 332–44. For the influence of contemporary research on the bonabo ape on the discussion about human aggression, see Frans DeWaal, *The Ape and the Sushi Master* (New York: Basic Books, 2001), 127–48.

69. See John R. Morss, *The Biologising of Childhood: Developmental Psychology and the Darwinian Myth* (London: Lawrence Erlbaum Associates, 1990). Also, Christine Atkinson, *Making Sense of Piaget: The Philosophical Roots* (London: Routledge and Kegan Paul, 1983), 171–94.

70. Ian Barbour, who explores the interface between science and religion, suggests a typology of relations between religion and science. The first he calls *conflict,* where two truth claims collide and only one can be true. That is the traditional view of the way in which the relationship between Darwinism and religion was first conceived, although the historical record is, in fact, far more complex. The second model he calls *independence,* where he argues that science and religion are separate worldviews, and relate to different enquiries. That is the strategy which has been adopted officially in most schools, although I would argue that the hidden curriculum usually supports a model of conflict. The model of independence is made possible by assuming an unbreachable wall between the "is" and the "ought," so that scientific discussions on human nature have no bearing on moral discussions of human aims and purposes. I argue that such a position is, in fact, both empirically impossible and morally problematic. The Darwinist narrative which is ultimately supported here is more in the spirit of Barbour's third and fourth model, *dialogue* and *integration,* where science and religion mutually impact one another. See Ian G. Barbour, *When Science Meets Religion* (New York: HarperCollins Publishers, 2000).

71. Darwin, *On the Origin of Species,* 106–7.

72. In anthropology, see Degler, 35–36. In economics, see, for example, Geoffrey Hodgson, *Economics and Evolution* (Cambridge: Polity Press, 1993), 123–38. In political science, see Mike Hawkins, *Social Darwinism in European and American Thought 1860–1945* (Cambridge: Cambridge University Press, 1997). In psychology, see Degler, 32–35. In sociology, see, for example, J. D. Y. Peel, *Herbert Spencer: The Evolution of a Sociologist* (London: Heinemann, 1971).

73. See Hawkins.

CHAPTER TWO. NATURE'S LESSONS

1. Stephen Jay Gould, "Kropotkin Was No Crackpot," in *Bully for Brontosaurus: Reflections in Natural History* (New York: W. W. Norton and Company, 1991), 333.

2. Peter Kropotkin, *Mutual Aid: A Factor of Evolution* (Boston, MA: Porter Sargent Publishers, 1914 edition), xiv.

3. For Spencer on Huxley's "Evolution and Ethics," see Herbert Spencer, "Evolutionary Ethics," *The Atheneum* 3432 (1893): 193–94, as quoted in James Paradis, "Evolution and Ethics in Its Victorian Context," in James Paradis and George C. Williams, *Evolution and Ethics* (Princeton: Princeton University Press, 1989), 44–46; for Dewey's reaction, see John Dewey, "Evolution and Ethics," Larry Hickman and Thomas Alexander, eds., *The Essential Dewey: Volume II* (Bloomington and Indianapolis: Indiana University Press, 1998), 225–35.

4. T. H. Huxley, "Evolution and Ethics," in Paradis and Williams, *Evolution and Ethics*, 109.

5. Ibid, 109–10.

6. Ibid, 139.

7. Paradis, 163.

8. Huxley, "Evolution and Ethics," 139–40.

9. "It is from neglect of these plain considerations that the fanatical individualism of our time attempts to apply the analogy of cosmic nature to society. Once more we have a misapplication of the stoical injunction to follow nature; the duties of the individual to the State are forgotten, and his tendencies to self-assertion are dignified by the name of rights" (Ibid, 140).

10. Lawrence A. Cremin, *The Transformation of the School: Progressivism in American Education 1876–1957* (New York: Vintage Books, 1961), 90–96.

11. For example, Spencer, "Social Statics," in Low-Beer, Ann, ed., *Herbert Spencer* (London: Collier-Macmillan, 1969), 31.

12. Huxley, "Evolution and Ethics," 139–40.

13. As described in explaining the Stoic notion of human nature. Ibid, 131–32.

14. Ibid, 141–42.

15. Ibid, 132.

16. "In the language of the Stoa, 'Nature' was a word of many meanings. There was the 'Nature' of the cosmos and the 'Nature' of man. In the latter, the animal 'nature', which man shares with a moiety of the living part of the cosmos, was distinguished from a higher 'nature.' Even in this higher nature there were grades of rank. The logical faculty is an instrument which may be turned to account for any purpose. The passions and the emotions are so closely tied to the lower nature that they may be considered to be pathological, rather than normal, phenomena. The one supreme, hegemonic, faculty, which constitutes the essential 'nature' of man, is most nearly represented by that which, in the language of a later philosophy, has been called the pure reason. It is this 'nature' which holds up the ideal of the supreme good and demands absolute submission of the will to its behests. It is this which commands all men to love one another, to return good for evil, to regard one another as fellow-citizens of one great state. Indeed, seeing that the progress towards perfection of a civilized state, or polity, depends on the obedience of its

members to these commands, the Stoics sometimes termed the pure reason the 'political' nature" (Ibid, 132–33).

17. Paradis, *T. H. Huxley: Man's Place in Nature*, 153–54.

18. Huxley, "Prolegomena," 101–3; on Huxley foreshadowing Freud, see Paradis, 152–54. For the idea in Freud's thought, see Sigmund Freud, *Civilization and Its Discontents* (New York: W. W. Norton and Company, 1961), 86–92.

19. T. H. Huxley, "Evolution and Ethics: Prolegomena," in Paradis and Williams, 67–69.

20. The metaphor of the gardener, and its interpretation, became a major theme in Darwinism after Huxley's introduction of the metaphor in his *Prolegemena*. As we shall see, Dewey used it in his response to Huxley's essay, and, in second-generation Darwinism, Rene Dubos utilizes it to articulate an anthropocentric environmental ethic. John Dewey, "Evolution and Ethics," Larry A. Hickman and Thomas M. Alexander, eds., *The Essential Dewey, Volume 2* (Bloomington and Indianapolis: University of Indiana Press, 1998), 225–35; Rene Dubos, "From Franciscan Conservation to Benedictine Stewardship," in *A God Within* (New York: Scribner, 1971).

21. Gould, "Kropotkin Was No Crackpot," 333.

22. Kropotkin, *Mutual Aid*, 1–6.

23. Kropotkin, in fact, argues that over the years Darwin withdrew somewhat from his theory of natural selection in general, and began to accept that Lamarckian evolution (inheritance of acquired characteristics through direct interrelation with the environment) was, in fact, a central factor in evolution. Kropotkin himself, although also arguing for mutual aid as a method of natural selection, also strongly argues for Lamarckian evolutionary mechanisms. See Peter Kropotkin, *Evolution and Environment* (Montreal: Black Rose Books, 1995), 118–34.

24. Kropotkin, *Mutual Aid*, vii–ix.

25. Charles Darwin, *On the Origin of Species* (Oxford: Oxford University Press, 1996; reproduction of second edition, 1860), 58.

26. Robert McIntosh, "Competition," in Evelyn Fox Keller and Elisabeth A. Lloyd, eds., *Keywords in Evolutionary Biology* (Cambridge, MA: Harvard University Press, 1992), 61.

27. Kropotkin, *Mutual Aid*, vii–xi.

28. "I indicated that warfare in Nature is chiefly *limited to struggle between different species* . . ." (Peter Kropotkin, *Ethics: Origin and Development* [Montreal: Black Rose Books, 1992], 14).

29. In *Mutual Aid*, first published in 1902, Kropotkin describes mutual aid as one of the factors of evolution, neglected by Darwinists. In *Ethics*, published posthumously twenty years later, Kropotkin calls mutual aid "the most predominant fact of nature." See Kropotkin, *Mutual Aid*, vii–xi; Kropotkin, *Ethics*, 14.

30. "Having its origin at the very beginnings of the evolution of the animal world, it is certainly an instinct as deeply seated in animals, low and high, as the instinct of maternal love; perhaps even deeper, because it is present in such animals as the molluscs, some insects, and most fishes, which hardly possess the maternal instinct at all" (Kropotkin, *Ethics*, 15).

31. Kropotkin, *Mutual Aid*, 293; also see Kropotkin, *Ethics*, 15.

32. Kropotkin, *Ethics*, 12.

33. "But if a scientist maintains that 'the only lesson which Nature gives to man is one of evil,' then he necessarily has to admit the existence of some other, extra-natural, or super-natural influence which inspires man with conceptions of 'supreme good,' and guides human development towards a higher goal. And in this way he nullifies his own attempt at explaining evolution by the action of natural forces only" (Ibid, 13).

34. Ibid, 13.

35. Gould argues that one should always be suspicious of views of nature which fit exactly with our attitudes about human society. The fact, for example, that Kropotkin's description of nature fits so well with his ethical hope for human society suggests that his science is problematic. Kropotkin is unreflective about this, simply relieved that nature and morality match up so well. Kropotkin, however, does argue that humans get their notion of good and bad from nature, suggesting that if nature functioned differently, human morality would be formulated differently (Kropotkin, *Ethics*, 16–17.) More sophisticated articulations of this is/ought issue are presented among contemporary Darwinists. See Gould, "Kropotkin Was No Crackpot," 338–39.

36. Kropotkin, *Mutual Aid*, 97.

37. Ibid, 97.

38. See, for example, Richard Lee, *The !Kung San: Men, Women, and Work in a Foraging Society* (Cambridge: Cambridge University Press, 1979).

39. Kropotkin, *Mutual Aid*, 292.

40. Kropotkin, *Mutual Aid*, 15.

41. The contradiction between moral inner group relations and immoral relations between groups is a central critique of communitarian ethics, which advocates a morality based on the community, and liberals have accused it of maintaining irrelevant moral distinctions between groups which undermine the moral unity of all of humanity. This issue is discussed in detail in Chapter Five, on education and Darwinism. See Joshua Cohen, ed., *For the Love of Country: Debating the Limits of Patriotism* (Boston: Beacon Press, 1996).

42. See, for example, Frank Salter, *On Genetic Interests: Family, Ethnicity and Humanity in an Age of Mass Migrations* (New York: Peter Lang Publishing, 2003).

43. "But in proportion as relations of equity and justice are solidly established in the human community, the ground is prepared for the further and the more general development of more refined relations, under which man understands and feels so well the bearings of his action on the whole of society that he refrains from offending others, even though he may have to renounce on that account the gratification of some of his own desires, and when he so fully identifies his feelings with those of others that he is ready to sacrifice his powers for their benefit without expecting anything in return" (Kropotkin, *Ethics*, 30).

44. Ibid, 30.

45. Kropotkin, *Mutual Aid*, 294. Kropotkin understands Spencer's ethics as sharing Hobbes' view of human nature: "Remaining true to Hobbes, he considers them loose aggregations of individuals who are strangers to one another, continually fighting and quarreling, and emerging from this chaotic state only after some

superior man, taking power into his hands, organizes social life" (Kropotkin, *Ethics,* 47).

46. Kropotkin, *Mutual Aid,* 299–300.
47. Ibid, 299–300.
48. Ibid, 294.
49. Ibid, 295.
50. Kropotkin does not deny individual competition as a factor in evolution. He just believes it has been grossly overstated, and that its complimentary mechanism, mutual aid, ignored. Ibid, xiv–xvii.
51. Kropotkin, *Ethics,* 19–20.
52. Kieran Egan, *Getting It Wrong from the Beginning: Our Progressivist Inheritance from Herbert Spencer, John Dewey, and Jean Piaget* (New Haven: Yale University Press, 2002).
53. Cremin, 91.
54. See, for example, Mike Hawkins, *Social Darwinism in European and American Thought* (Cambridge: Cambridge University Press, 1997).
55. J. D. Y. Peel, *Herbert Spencer: The Evolution of a Sociologist* (London: Heinemann, 1971), 141–53; Hawkins, however, sees Spencer's philosophy as a synthesis of Lamarckian and Darwinian perspectives. See Hawkins, 87–88.
56. Herbert Spencer, "Moral Education," in *Education: Intellectual, Moral and Physical* (London: Watts, 1941 [1861]), 103.
57. Herbert Spencer, "Social Statics," as quoted in Low-Beer, 153.
58. On the ornamental versus the useful, see Spencer, "What Knowledge Is of Most Worth," in *Education: Intellectual, Moral and Physical,* 16; on appearance versus function, see Ibid, 1–2; on conventional versus intrinsic value, see Ibid, 11.
59. Ibid, 11.
60. In describing current educational fashion, Spencer mocks the emphasis on external appearances, rather than intrinsic value determined by its utility: "We are none of us content with quietly unfolding our own individualities to the full in all directions; but have a restless craving to impress our individualities upon others, and in some way subordinate them. And this it is which determines the character of our education. Not what knowledge is of most real worth, is the consideration; but what will bring most applause, honour, respect—what will most conduce to social position and influence—what will be most imposing. As, throughout life, not what we are, but what we shall be that, is the question; so in education, the question is, not the intrinsic value of knowledge, so much as its extrinsic effects on others. And this being our dominant idea, direct utility is scarcely more regarded than by the barbarian when filing his teeth and staining his nails" (Ibid, 4).
61. For example, "Society is made up of individuals; all that is done in society is done by the combined actions of individuals; and therefore, in individual actions only can be found the solutions of social phenomena. But the actions of individuals depend on the laws of their natures; and their actions cannot be understood until these laws are understood" (Ibid, 34).
62. Spencer uses the terms 'instincts,' 'emotions,' 'feelings' and 'sensations' as synonyms. Referring to 'instincts,' see Spencer, "Moral Education," 127, or Spencer, "Physical Education," *Education: Intellectual, Moral and Physical,* 176. Referring

to 'emotions' see Herbert Spencer, *Facts and Comments* (New York: D. Appleton and Co., 1902), as quoted in Low-Beer, 157. Referring to 'feelings,' see Spencer, "Moral Education," 133. Referring to 'sensations,' see Spencer, "Physical Education," 152.

63. Spencer, "What Knowledge Is of Most Worth," 12. Spencer's particular example does not hold up to present knowledge of behavior among human babies, but his point that there is instinctual behavior, and that human beings are not blank slates, is central to the argument of other Darwinians, particularly Dewey in first-generation Darwinism, and Midgley in the second generation.

64. Spencer, *Facts and Comments,* quoted in Low-Beer, 84. The emphasis on feelings as a central component of human nature supports a similar theme in Darwin, as well.

65. Spencer, "Physical Education," 152.

66. For example, Spencer, "What Knowledge Is of Most Worth," 30.

67. Ibid, 30.

68. "When a father, acting on the false dogmas, adopted without examination, has alienated his sons, driven them into rebellion by his harsh treatment, ruined them, and made himself miserable; he might reflect that the study of Ethology would have been worth pursuing, even at the cost of knowing nothing about Aeschylus . . . when she is prostrate under the pangs of combined grief and remorse; it is but small consolation that she can read Dante in the original" (Spencer, "What Knowledge Is of Most Worth," 29).

69. For example, Spencer, *The Study of Sociology,* as quoted in Low-Beer, 34.

70. For centrality of instincts, see Spencer, "What Knowledge Is of Most Worth," 14. For centrality of science, see Ibid, 46.

71. Cremin, 93.

72. Spencer, *Facts and Comments,* in Low-Beer, 88.

73. "Education has for its object the formation of character. To curb restive propensities, to awaken dormant sentiments, to strengthen the perceptions and cultivate the tastes, to encourage this feeling and repress that, so as finally to develop the child into a man of well-proportioned and harmonious nature—this is alike the aim of parent and teacher." Spencer, *Social Statics,* in Low-Beer, 126.

74. Spencer, "Physical Education," 161, 165, 167.

75. Spencer, "What Knowledge Is of Most Worth?," 49.

76. Ibid, 49.

77. "Thus:—Society is made up of individuals; all that is done in society is done by the combined actions of individuals; and therefore, in individual actions only can be found the solutions of social phenomena. But the actions of individuals depend on the laws of their natures; and their actions cannot be understood until these laws are understood. These laws, however, when reduced to their simplest expressions, prove to be corollaries from the laws of body and mind in general. Hence it follows, that biology and psychology are indispensable as interpreters of sociology" (Ibid, 34).

78. Peel, vii–viii.

79. "Unexpected though the assertion may be, it is nevertheless true, that the highest Art of every kind is based on Science—that without Science there can be

neither perfect production nor full appreciation" (Spencer, "What Knowledge Is of Most Worth?," 37).

80. "The reply is—Philosophy may still properly be the title retained for knowledge of the highest generality. Science means merely the family of the Sciences—stands for nothing more than the sum of knowledge formed of their contributions; and ignores the knowledge constituted by the *fusion* of these contributions into a whole. As usage has defined it, Science consists of truths existing more or less separated and does not recognize these truths as entirely integrated . . ." (Herbert Spencer, *First Principles* [New York: D. Appleton and Co., 1862], in Low-Beer, 57).

81. Spencer, "Intellectual Education," 80.

82. Spencer, "What Knowledge Is of Most Worth?," 13.

83. Ibid, 12–13.

84. Ibid, 108.

85. "It is the peculiarity of these penalties, if we must so call them, that they are simply the *unavoidable consequences* of the deeds which they follow: they are nothing more than the *inevitable reactions* entailed by the child's actions" (Ibid, 107).

86. Ibid, 115.

87. See Low-Beer, 118; and, for examples, Spencer, *The Principles of Ethics Volume I*, in Low-Beer, 136–37.

88. Spencer, "Intellectual Education," 63. Much has been made of the extended period of childhood among humans. The fact has often been used to show how learning makes human nature particularly elastic—that, because of the long period of human childhood, human beings are a unique learning animal, and therefore capable of developing in virtually infinite ways. Gould particularly makes this point in second-generation Darwinism.

89. For example: "We have fallen upon evil times, in which it has come to be an accepted doctrine that part of the responsibilities are to be discharged not by parents but by the public—a part which is gradually becoming a larger part and threatens to become the whole" (Spencer, *The Principles of Ethics Volume I*, in Low-Beer, 137).

90. Spencer, "What Knowledge Is of Most Worth?," 46.

91. Spencer, "Intellectual Education," 71.

92. "As a final test by which to judge any plan of culture should come the question, Does it create a pleasurable excitement in the pupils" (Spencer, "Intellectual Education," 73).

93. "Not by authority is your sway to be obtained; neither by reasoning; but by inducement. Show in all your conduct that you are thoroughly your child's friend, and there is nothing that you may not lead him to" (Spencer, *Social Statics*, in Low-Beer, 129).

94. An encapsulated view of the educational revolution that Spencer intended: "What now is the common characteristic of these several changes? Is it not an increasing conformity to the methods of Nature? The relinquishment of early forcing, against which Nature rebels, and the leaving of the first years for exercise of the limbs and senses, show this. The superseding of rote-learnt lessons by lessons orally and experimentally given, like those of the field and playground, shows this. And above all, this tendency is shown in the variously-directed efforts to present

knowledge in attractive forms, and so to make the acquirement of it pleasurable" (Spencer, "Intellectual Education," 60). For examples of Spencer's egalitarian attitude towards girls, see Spencer, "What Knowledge Is of Most Worth?," 13; Spencer, "Physical Education", 158–61.

95. Spencer, "Moral Education," 133.

96. See, for example, Huxley, "A Liberal Education and Where to Find It," 208–14.

97. For example: "Such a one and no other, I conceive, has had a liberal education; for he is, as completely as a man can be, in harmony with nature. He will make the best of her, and she of him" (in Huxley, "A Liberal Education and Where to Find It," 211). About living according to nature's laws: "Well, what I mean by Education is learning the rules of this mighty game. In other words, education is the instruction of the intellect in the laws of nature, under which name I include not merely things and their forces, but men and their ways; and the fashioning of the affections and of the will into an earnest and loving desire to move in harmony with those laws" (Ibid, 209).

98. Alan Barr, ed., *The Major Prose of Thomas Henry Huxley* (Athens, GA: University of Georgia Press, 1997), 205.

99. Huxley, "A Liberal Education and Where to Find It," 210–11.

100. Ibid, 212.

101. For example, in discussing the causes of poverty, Huxley argues that much of its root is in ignorance of the laws of nature—poor diet, unsanitary conditions, mistreatment of illness. Knowing nature's laws can avoid much suffering in the world. Ibid, 213.

102. For example: "In whichever way we look at the matter, morality is based on feeling, not on reason, though reason alone is competent to trace out the effects of our actions and thereby dictate conduct. . . . The moral law, like the laws of physical nature, rests in the long run upon instinctive intuitions, and is neither more nor less 'innate' and 'necessary' than they are" (T. H. Huxley, "Hume," in Cyril Bibby, ed., *T. H. Huxley on Education* [Cambridge: Cambridge University Press, 1971], 178–79).

103. Huxley, "Hume," in Bibby, 173. On contemporary language and Darwinism, see Steven Pinker, *The Language Instinct: How the Mind Creates Language* (New York: Harper Collins Books, 1994).

104. Huxley, "A Liberal Education and Where to Find It," 210.

105. Ibid, 210.

106. Huxley, "Scientific Education," in Bibby, 104.

107. Huxley, "A Liberal Education and Where to Find It," 210.

108. Ibid, 211.

109. Ibid, 211.

110. Ibid, 211.

111. On science and rationality, see, for example, "We falsely pretend to be the inheritors of their [Greek] culture, unless we are penetrated, as the best minds among them were, with an unhesitating faith that the free employment of reason, in accordance with scientific method, is the sole method of reaching truth" (Huxley, "Science and Culture," 234). On science and progress, see, for example, Ibid, 236.

112. Huxley, "Evolution and Ethics," 132–33.

113. Huxley, "A Liberal Education and Where to Find It," 214.

114. "The mental power which will be of most importance in your daily life will be the power of seeing things as they are without regard to authority; and of drawing accurate general conclusions from particular facts. But at school and at college you shall know of no source of truth but authority; nor exercise your reasoning faculty upon anything but deduction from that which is laid down by authority" (Ibid, 216).

115. Ibid, 215.

116. Huxley, "Science and Culture," in Barr, ed., *The Major Prose of T. H. Huxley*, 234.

117. T. H. Huxley, "Technical Education," in Bibby, 159.

118. See, for example, Huxley, "Address on the Behalf of the National Association for the Promotion of Technical Education," in Bibby, 199–204.

119. "But if the classics were taught as they might be taught—if boys and girls were instructed in Greek and Latin, not merely as languages, but as illustrations of philological science; if a vivid picture of life on the shores of the Mediterranean two thousand years ago were imprinted on the minds of scholars; if ancient history were taught, not as a wary series of feuds and fights, but traced to its causes in such men placed under such conditions; if, lastly, the study of the classical books were followed in such a manner as to impress boys with their beauties, and with the grand simplicity of their statement of the everlasting problems of human life, instead of with their verbal and grammatical peculiarities; I still think it as little proper that they should form the basis of a liberal education for our contemporaries, as I should think it fitting to make that sort of paleontology with which I am familiar the back-bone of modern education" (Huxley, "A Liberal Education and Where to Find It," 217).

120. Ibid, 66.

121. Progressive education's demands for a 'practical' education have been often accused of leading to a crass utilitarianism in order to serve the technical needs of the economic and social status quo. This theme emerges particularly in the critique of John Dewey's educational philosophy. See Alan Ryan, *John Dewey and the High Tide of American Liberalism* (New York: W. W. Norton and Company, 1995), 345–47.

122. For example: "I judge the value of human pursuits by their bearing upon human interests; in other words, by their utility . . ." (T. H. Huxley, "On the Study of Biology," in Bibby, 150); or, "The great end of life is not knowledge but action" (Huxley, "Technical Education," in Bibby, 162).

123. "There are other forms of culture beside physical science; and I should be profoundly sorry to see the fact forgotten, or even to observe a tendency to starve, or cripple, literary, or aesthetic, culture for the sake of science" (Huxley, "Scientific Education," in Bibby, 101).

124. Ibid, 105.

125. Huxley, "Science and Culture," in Barr, 236.

126. Huxley, "Emancipation: Black and White," in Bibby, 69.

127. Ibid, 69.

128. Peter Kropotkin, "Brain Work and Manual Work," in *Fields, Factories and Workshops* (London: T. Nelson, 1912), 3.
129. Ibid, 1–2.
130. Ibid, 14–16.
131. Ibid, 1, 15.
132. Ibid, 1–2.
133. Ibid, 15.
134. Ibid, 10.
135. For example, Kropotkin, *Ethics,* 45. Group selection had been largely discarded in evolutionary theory, but has recently made somewhat of a comeback. For a defense of group selection theory in contemporary evolutionary theory, see Elliott Sober and David Sloan Wilson, *Unto Others: The Evolution and Psychology of Unselfish Behavior* (Cambridge, MA: Harvard University Press, 1998).
136. "That such a science is possible lies beyond any reasonable doubt. If the study of Nature has yielded the elements of a philosophy which embraces the life of the Cosmos, the evolution of living beings, the laws of physical activity and the development of society, it must also be able to give us the rational origin and the sources of moral feelings. And it must be able to show us where lie the forces that are able to elevate the moral feeling to an always greater height and purity" (Kropotkin, *Ethics,* 5).
137. Ibid, 4.
138. Kropotkin, "Brain Work and Manual Work," 16.
139. Ibid, 16.
140. Ibid, 15.
141. Ibid, 16.
142. "How much better the historian and the sociologist would understand humanity if they knew it, not in books only, not in a few of its representatives, but as a while, in its daily life, daily work, and daily affairs! How much more medicine would trust to hygiene, and how much less to prescriptions, if the young doctors were the nurses of the sick and the nurses received the education of the doctors of our time!" (Ibid, 16).
143. Ibid, 7.
144. Ibid, 14.
145. "Of course, we have a number of cases in which the discovery, or the invention, was a mere application of a scientific law (cases, like the discovery of the planet Neptune), but in the immense majority of cases the discovery, or the invention, is unscientific to begin with. It belongs much more to the domain of art—art taking the precedence over science" (Ibid, 15).
146. Ibid, 6–7.
147. Ibid, 7.
148. Ibid, 7.
149. Ibid, 8.
150. Spencer subscribed to the widespread notion of the nineteenth century that ontogeny recapitulates phylogeny—that the development of the embryo reenacted the stages of evolutionary development. Spencer expanded that notion, believing that childhood development, likewise, went through the evolutionary stages of

human culture, so that children begin as wild beasts and savages. His portrayal of early childhood is a window into his understanding of 'primitive' culture: "Those respects in which a child requires restrain, are just the respects in which he is taking after the aboriginal man. The selfish squabbles of the nursery, the persecution of the playground, the lyings and petty thefts, the rough treatment of inferior creatures, the propensity to destroy—all these imply that tendency to pursue gratification at the expense of other beings which qualified man for the wilderness, and which disqualifies him for civilized life" (in Spencer, *Social Statics,* quoted in Low-Beer, 31).

CHAPTER THREE. DEWEY'S DARWNISM

1. On communitarian understandings of Dewey, see, for example, Steven C. Rockefeller, *John Dewey: Religious Faith and Democratic Humanism* (New York: Columbia University Press, 1991). Even Alan Ryan, along with Rockefeller a sensitive reader of Dewey, is dismissive of evolution's ability to be a significant factor in explaining human motives, although acknowledging its centrality for Dewey. Ryan essentially argues that Dewey is relevant in spite of his evolutionary perspective, and not because of it. See Alan Ryan, *John Dewey and the High Tide of American Liberalism* (New York: W. W. Norton and Company, 1995), 130. For an attempt to frame Dewey as an Aristotelian, see Terry Hoy, *Toward a Naturalistic Political Theory: Aristotle, Hume, Dewey, Evolutionary Biology, and Deep Ecology* (Westport, CT: Praeger, 2000).

2. For his initial assertion of his Darwinism, see John Dewey, "The Influence of Darwinism on Philosophy," in Larry Hickman and Thomas Alexander, eds., *The Essential Dewey: Volume I* (Bloomington and Indianapolis: Indiana University Press, 1998), 39–45.

3. "The conceptions that had reigned in the philosophy of nature and knowledge for two thousand years, the conception that had become the familiar furniture of the mind, rested on the assumption of the superiority of the fixed and final; they rested upon treating change and origin as signs of defect and unreality. In laying hands upon the sacred ark of absolute permanency, in treating the forms that had been regarded as types of fixity and perfection as originating and passing away, the *On the Origin of Species* introduced a mode of thinking that in the end was bound to transform the logic of knowledge, and hence the treatment of morals, politics and religion" (Ibid, 39).

4. Ibid, 40–41.
5. Ibid, 40–41.
6. Ibid, 39–40.
7. Ibid, 43.
8. See John Dewey, "Anti-Naturalism in Extremis," in *The Essential Dewey: Volume I,* especially 165–66.
9. A similar point is made about Dewey by Alven Neiman, "Pragmatism: The Aims of Education and the Meaning of Life," in Wendy Kohli, ed., *Critical Conversations in Philosophy of Education* (London: Routledge, 1995), 56–72.
10. Dewey, "The Influence of Darwinism on Philosophy," 43.
11. Ibid, 43.

12. "Finally, the new logic introduces responsibility into the intellectual life. To idealize and rationalize the universe at large is after all a confession of inability to master the course of things that specifically concern us. As long as mankind suffered from this impotency, it naturally shifted a burden of responsibility that it could not carry over to the more competent shoulders of the transcendent cause. But if insight into specific conditions of value and into specific consequences of ideas is possible, philosophy must in time become a method of locating and interpreting the more serious of the conflicts that occur in life, and a method of projecting ways for dealing with them: a method of moral and political diagnosis and prognosis" (Ibid, 44).

13. Darwin most often understood evolution as a radiating tree with no clear hierarchy. See Charles Darwin, *On the Origin of Species* (Oxford: Oxford University Press, 1996), 106–7.

14. Ibid, 8–11.

15. For an excellent discussion of Dewey's concept of growth, see Daniel Pekarsky, "Dewey's Conception of Growth Reconsidered," *Educational Theory* 40, No. 3 (1990): 283–94.

16. John Dewey, *Democracy and Education* (New York: The Free Press, 1944), 41–47.

17. Ibid, 44–45.

18. Pekarsky, "Dewey's Conception of Growth Reconsidered," 285–86.

19. For example, "Since the latter accept without discount and qualification facts that are authenticated by careful and thorough inquiry, they recognize in their full force observed facts that disclose the differences existing between man and other animals, as well as the strands of continuity that are discovered in scientific investigation. The idea that there is anything in naturalism that prevents acknowledgment of differential traits in their full significance, or that compels their 'reduction' to traits characteristic of worms, clams, cats, or monkeys has no foundation" (John Dewey, "Anti-Naturalism in Extremis," in *The Essential Dewey, Volume I*, 168).

20. See the discussion of Hillary Putnam and Richard Rorty in the context of James's model in Russell B. Goodman, "Introduction," in Goodman, ed., *Pragmatism* (London: Routledge, 1995), 6–11.

21. Dewey, for example, in describing the animal legacy in human behavior, writes: "But man as man still has the dumb pluck of the animal. He has endurance, hope, curiosity, eagerness, love of action. These traits belong to him by structure, not by taking thought." (John Dewey, *Human Nature and Conduct* [New York: Henry Holt, 1922], 200).

22. See Rockefeller, 365.

23. "Observation of a child, even a young baby, will convince the observer that a normal human being when awake is engaged in activity; he is a reservoir of energy that is continually overflowing. The organism moves, reaches, handles, pulls, pounds, tears, molds, crumples, looks, listens, etc." (John Dewey and James H. Tufts, *Ethics*, ed. Jo Anne Boyston [Carbondale: Southern Illinois University Press, 1976], 343).

24. See Dewey, *Human Nature and Conduct*, 75.

25. Ibid, 136.

26. Ibid, 4.

27. See, for example, Dewey's comment on the life of a wolf: "The fact is the wolf asserted himself as a wolf. It was not mere life he wished, but the life of the wolf. No agent can draw this distinction between desire for mere life and desire for happy life for himself; and no more can the spectator intelligently draw it for another" (John Dewey, "Evolution and Ethics," in *The Essential Dewey, Volume II*, 230).

28. "Dewey's experimental empiricism approaches moral problems in a fashion quite similar to that found in classical Greek philosophy in the sense that he emphasizes the central significance of need, desire, and satisfaction in human behavior. However, his biological and social orientation gives his treatment of these factors a distinctive character. He views the human being as a creature of needs and wants interacting with a problematic world" (Rockefeller, 408).

29. Dewey, *Human Nature and Conduct*, 136.

30. Dewey, *Democracy and Education*, 44–45.

31. Ibid, 114.

32. Ibid, 207–8.

33. Ibid, 80.

34. Ibid, 83.

35. Ibid, 80.

36. "Habits as organized activities are secondary and acquired, not native and original. They are outgrowths of unlearned activities which are part of man's endowment at birth" (Ibid, 66).

37. Dewey, Ibid, 44.

38. Dewey, Ibid, 124.

39. Dewey, *Democracy and Education*, 74–75.

40. Dewey, *Human Nature and Conduct*, 124.

41. Dewey, Ibid, 21.

42. John Dewey, "The Reflex Arc Concept in Psychology," in *The Essential Dewey, Volume I*, 3–10.

43. Dewey, *Ethics*, 342. A similar description of the relationship between instinct and thought appears in Dewey, *Human Nature and Conduct*, 207: "The problem of morals is the problem of desire and intelligence. What is to be done with these facts of disharmony and conflict? After we have discovered the place and consequences of conflict in nature, we have still to discover its place and working in human need and thought. What is its office, its function, its *possibility*, or use? In general, the answer is simple. Conflict is the gadfly of thought. It stirs us to observation and memory. It instigates to inventions. It shocks us out of sheep-like passivity, and sets us at noting and contriving."

44. Dewey, *Human Nature and Conduct*, 216.

45. Dewey, *Democracy and Education*, 75.

46. In *Experience and Education*, Dewey tries to distance himself from those progressive educators who understood progressive education as implying that guidance of the child is no longer needed, and socialization as an anathema; see John Dewey, *Experience and Education* (New York: Collier Books, 1963).

47. "If different situations were wholly unlike one another, nothing could be learned from one which would be of any avail in any other. But having like points, experience carries over from one to another, and experience is intellectually cumulative" (Dewey and Tufts, *Ethics*, 335).

48. Dewey, *Human Nature and Conduct*, 125.

49. "But (a) nothing is more immediate and seemingly sure of itself than inveterate prejudice. The morals of a class, clique, or race when brought in contact with those of other races and peoples, are usually so sure of the rectitude of their own judgments of good and bad that they are narrow and give rise to misunderstanding and hostility. (b) A judgment which is adequate under ordinary circumstance may go far astray under changed conditions" (Dewey and Tufts, *Ethics*, 331).

50. For example, "There was a political slant to this denial of the native and a priori, this magnifying of the accomplishments of acquired experience. It held out a prospect of continuous development, of improvement without end. Thus writers like Helvetius made the idea of the compete malleability of a human nature which originally is wholly empty and passive, the basis for asserting the omnipotence of education to shape human society, and the ground of proclaiming the infinite perfectibility of mankind" (Dewey, *Human Nature and Conduct*, 76).

51. Dewey, Ibid, 77.

52. "For impulse when it asserts itself deliberately against an existing custom is the beginning of individuality in mind. This beginning is developed and consolidated in the observations, judgments, inventions which try to transform the environment so that a variant deviating impulse may itself in turn become incarnated in objective habit" (Dewey, *Human Nature and Conduct*, 62).

53. Dewey, "Evolution and Ethics," in *The Essential Dewey, Volume II*, 227.

54. Ibid, 228.

55. Ibid, 228.

56. A surprisingly eloquent Deweyism: "From a social standpoint, dependence denotes a power rather than a weakness; it involves interdependence. There is always a danger that increased personal independence will decrease the social capacity of an individual. In making him more self-reliant, it may make him more self-sufficient; it may lead to aloofness and indifference. It often makes an individual so insensitive in his relations to others as to develop an illusion of being really able to stand and act alone—an unnamed form of insanity which is responsible for a large part of the remediable suffering of the world" (Dewey, *Democracy and Education*, 44).

57. Dewey and Tufts, *Ethics*, 350.

58. For example: "Where egotism is not made the measure of reality and value, we are citizens of this vast world beyond ourselves, and any intense realization of its presence with and in us brings a peculiarly satisfying sense of unity in itself and with ourselves" (John Dewey, *Art as Experience* [New York: Minton, Balch and Co., 1934]); and again: "Militant atheism is also affected by lack of natural piety. The ties binding man to nature that poets have always celebrated are passed over lightly. The attitude taken is often that of man living in an indifferent and hostile world and issuing blasts of defiance. A religious attitude, however, needs the sense of a connection of man, in the way of both dependence and support, with the enveloping world

that the imagination feels is a universe" (John Dewey, *A Common Faith*,[New Haven: Yale University Press, 1972], 55).

59. "From the standpoint of its *definite* aim any act is petty in comparison with the totality of natural events. What is accomplished directly as the outcome of a turn which our action gives the course of events is infinitesimal in comparison with their total sweep. Only an illusion of conceit persuades us that cosmic difference hangs upon even our wisest and most strenuous effort. Yet discontent with this limitation is as unreasonable as relying upon an illusion of external importance to keep ourselves going. In a genuine sense every act is already possessed of infinite import. The little part of the scheme of affairs which is modifiable by our efforts is continuous with the rest of the world. The boundaries of our garden plot join it to the world of our neighbors and our neighbors' neighbors. That small effort which we can put forth is in turn connected with an infinity of events that sustain and support it. The consciousness of this encompassing infinity of connections is ideal. When a sense of the infinite reach of an act physically occurring in a small point of space and occupying a petty instant of time comes home to us, the *meaning* of a present act is seen to be vast, immeasurable, unthinkable. This ideal is not a goal to be attained. It is a significance to be felt, appreciated. Though consciousness of it cannot become intellectualized (identified in objects of a distinct character) yet emotional appreciation of it is won only by those willing to think" (Dewey, *Human Nature and Conduct*, 180).

60. "A person entirely lacking in sympathetic response might have a keen calculating intellect, but he would have no spontaneous sense of the claims of others for satisfaction of their desires. A person of narrow sympathy is of necessity a person of confined outlook upon the scene of human good. . . . It is sympathy which carries thought out beyond the self and which extends its scope till it approaches the universal as its limit. It is sympathy which saves consideration of consequences from degenerating into mere calculation . . ." (Dewey, *Ethics*, 333).

61. See, for example, Steven Rockefeller, *John Dewey: Religious Faith and Democratic Humanism* (1991); Alan Ryan, *John Dewey and the High Tide of American Liberalism* (1995); and Robert Westbrook, *John Dewey and American Democracy* (Ithaca: Cornell University Press, 1991). Each, in his own way, places Dewey into communitarian categories. Communitarian political theory emerged in the 1980s and held that the individual is constituted by his/her community, and is not an atomized entity existing independently of relationships. This is both an ontological point (human life cannot be understood except from within the context of its historical or social community) as well as a normative one (we should pursue our life's aims on the basis of our historical and/or social situation).

62. See, for example, Alistair MacIntyre, *After Virtue* (Notre Dame: University of Notre Dame Press, 1981), 203.

63. Ibid, 204–7.

64. Ryan, *John Dewey and the High Tide of American Liberalism*, 367–68.

65. For a clear explanation of Dewey's concept of ends-in-view, see Murray G. Murphy, "Introduction," in Dewey, *Human Nature and Conduct*, xviii.

66. Dewey, *Ethics*, 351.

67. Dewey, *Human Nature and Conduct*, 227.

68. *Democracy and Education*, 87, as cited and expanded upon in Pekarsky, "Dewey's Conception of Growth Reconsidered," 284. On 'the great community,' see John Dewey, "Search for the Great Community," in *The Essential Dewey, Volume I*, 307.

69. See such a sentiment in Dewey, "Search for the Great Community," 297.

70. John Dewey, "Creative Democracy: The Task Before Us," in *The Essential Dewey, Volume I*, 342.

71. On the rationale of people who emphasize only freedom *from* governmental involvement, see Dewey, *Human Nature and Conduct*, 210. Dewey's distinction between freedom *from* and freedom *to* foreshadows Isaiah Berlin's famous distinction between negative and positive liberty. See Isaiah Berlin, *Four Essays on Liberty* (London: Oxford University Press, 1975). Dewey's call to recast a balance between the two, which has been slanted toward freedom *from*, clearly foreshadows the communitarian critique of the classic liberal rights position, and is one of the main reasons Dewey was adopted by communitarians.

72. See the discussion in Ryan, comparing Dewey to Habermas (Ryan, *John Dewey and the High Tide of American Liberalism*, 357).

73. "A genuinely democratic faith in peace is faith in the possibility of conducting disputes, controversies and conflicts as cooperative undertakings in which both parties learn by giving the other a chance to express itself, instead of having one party conquer by forceful suppression of the other—a suppression which is none the less one of violence when it takes place by psychological means of ridicule, abuse, intimidation, instead of by overt imprisonment or in concentration camps" (Dewey, "Creative Democracy—The Task Before Us," in *The Essential Dewey, Volume I*, 342).

74. Dewey, "Anti-Naturalism in Extremis," 170.

75. Dewey and Tuft, *Ethics*, 349.

76. For example, Dewey, "Search for the Great Community," 301–2.

77. Dewey, *Democracy and Education*, 317.

78. See also Daniel Pekarsky, "Burglars, Robber Barons, and the Good Life," *Educational Theory* 44, No. 1 (1991): 73–74.

79. Dewey, "Creative Democracy—The Task Before Us," 341; also Dewey, *Democracy and Education*, 84–85.

80. Dewey and Tuft, *Ethics*, 349.

81. Ibid, 349.

82. Dewey, *Democracy and Education*, 49–50; Murray G. Murphey, "Introduction," *Human Nature and Conduct*, xiv.

83. John Dewey, *Experience and Education* (New York: Collier Books, 1963), 36.

84. "If the standard of morals is low it is because the education given by the interaction of the individual with his social environment is defective. Of what avail is it to preach unassuming simplicity and contentment of life when communal admiration goes to the man who "succeeds"—who makes himself conspicuous and envied because of command of money and other forms of power?" Dewey, *Human Nature and Conduct*, 219.

85. For his response to such critics, see, for example, Dewey, "Anti-Naturalism in Extremis," 169.

86. Dewey, *Democracy and Education*, 112.

87. For an explication of Dewey's interpretation of Rousseau in this context, see Dewey, *Democracy and Education*, 112–13.
88. Ibid, 114.
89. Ibid, 116.
90. Ibid, 114.
91. Ibid, 114–15.
92. Dewey's *Experience and Education* was written in order to clarify his thought, and to distance himself from what he perceived as the misinterpretation of his philosophy by so-called 'progressive education,' which could be seen as more Rousseauian than Deweyan. See John Dewey, "Traditional vs. Progressive Education," in Dewey, *Experience and Education*, 17–23.
93. See short discussion in Ryan, *John Dewey and the High Tide of American Liberalism*, 341–42.
94. Dewey, *Experience and Education*, 22; also, see Alan Ryan, *Liberal Anxieties and Liberal Education* (New York: Hill and Wang, 1998), 113–14.
95. Ibid, 113–14.
96. Dewey, *Human Nature and Conduct*, 25.
97. Dewey, *Experience and Education*, 17–19.
98. Dewey, *Human Nature and Conduct*, 43.
99. Ibid, 29–30.
100. John Dewey, "Moral Principles in Education," in *The Essential Dewey, Volume I*, 248.
101. See Rockefeller, 427–28, and Ryan, *John Dewey and the High Tide of American Liberalism*, 137–48.
102. John Dewey, "Self-Realization as the Moral Ideal," in John Dewey, *Early Works*, 4:50, as quoted in Steven C. Rockefeller, 427.
103. Ibid, 427–28.
104. John Dewey, "How We Think," in *Middle Works, Volume 6*, 215, as quoted in Ryan, *John Dewey and the High Tide of American Liberalism*, 143.
105. Ryan, *John Dewey and the High Tide of American Liberalism*, 367–68.
106. Ibid, 172.
107. Ibid, 172.
108. For a similar point, see Ibid, 364.

CHAPTER FOUR. MARY MIDGLEY AND THE ECOLOGICAL *TELOS*

1. Peter J. Bowler, *The Eclipse of Darwinism* (Baltimore: The Johns Hopkins University Press, 1983), 182–226.
2. Peter J. Bowler, *The Norton History of the Environmental Sciences* (New York: W. W. Norton and Company, 1992), 445–73.
3. For a full historical analysis of the waning and rebirth of Darwinism in America, see Carl Degler, *In Search of Human Nature: The Decline and Revival of Darwinism in American Social Thought* (New York: Oxford University Press, 1991).
4. For an extended discussion of the reasons for the shift, see Degler, 187–211.
5. Ibid, 59–83.

6. Ibid, 161–211.
7. Ibid, 202–4.
8. See discussion in Kenan Malik, *Man, Beast and Zombie* (London: Weidenfeld and Nicolson, 2000), 115–47.
9. Degler, 215–44.
10. E. O. Wilson, *Sociobiology: The New Synthesis* (Cambridge, MA: Harvard University Press, 1975); Richard Dawkins, *The Selfish Gene* (Oxford: Oxford University Press, 1989 [1976]).
11. Andrew Brown, *The Darwin Wars: The Scientific Battle for the Soul of Man* (London: Simon and Schuster, 1999).
12. " . . . after reading the ethologists—especially Tinbergen and Lorenz—I thought that they were making new and very useful contributions to the enquiry about the difficult subject of human motives. I thought that these contributions ought to be digested, not spat out in disgust. They could (I believed) bring much-needed light to moral philosophy, which was my official academic business. They were surely relevant to philosophical questions about the relation between body and mind, between motivation and rationality, between humanity and nature. They could help to answer queries on these topics already launched by psychoanalytic thinkers, queries which I also considered important. . . . I was astonished, then, at the social scientists' shocked and defensive response, their determination to retreat from the whole subject behind the species barrier" (Mary Midgley, *Beast and Man* [London: Routledge, 1995, second edition], xv).
13. Ibid, xiv.
14. Midgley, *Beast and Man*, 19–22.
15. Midgley, *Beast and Man*, xxxix.
16. For the original coining of the phrase, see Gilbert Ryle, *The Concept of Mind* (London: Penguin Books, 1949), 17.
17. For example, Mary Midgley, *The Ethical Primate* (London: Routledge, 1994), 7–9.
18. Ibid, 9.
19. Midgley, *Beast and Man*, 4.
20. "Some people seem to think that the suddenness of 'punctuated evolution' [Gould's theory which challenged conventional Darwinist theory] can isolate us from our past—that a 'hopeful monster' may have appeared which bore all the distinctive marks of modern humanity and no inconvenient heritage from previous ancestors" (Midgley, *The Ethical Primate*, 15).
21. See E. Allen et al., "Letter." *The New York Review of Books*, Volume 13, November, 1975, 182; Stephen Jay Gould, *An Urchin in the Storm* (New York: W.W. Norton and Company, 1987), 112–15.
22. Midgley, *Beast and Man*, xvii.
23. Mary Midgley, "Gene-Juggling," *Philosophy* 54 (1979): 439–58.
24. "Selection does not work by cutthroat competition between individuals, but by favoring whatever behavior is useful to the group. People with crude notions of "Darwinism" make an intriguing blunder here. They confuse the mere *fact of competing*, that is, of needing to share out a resource, with the *motive of competitiveness* or readiness to quarrel. Where creatures are competing (as a fact), their success

will be decided by whatever tendencies they have that best help their predicament. These need not be quarrelsome tendencies at all" (Midgley, *Beast and Man,* 132).

25. Eliot Sober and David Sloan Wilson, in their work on group selection, show how four possibilities of interaction can exist between evolutionary mechanisms and psychological motives/behaviors: evolutionary selfishness can lead to both psychological egoism and altruism; but they hold that an evolutionary altruism is also possible, which can also lead to both an egoistic or altruistic psychology. See Eliot Sober and David Sloan Wilson, *Unto Others: The Evolution and Psychology of Unselfish Behavior* (Cambridge, MA: Harvard University Press, 1998), 202–8.

26. Quoted in Mary Midgley, *Science and Poetry,* (New York and London: Routledge, 2001), 194.

27. "We must avoid the Panglossian confidence with which some evolutionists today declare everything to have a function. Some features of organisms really are passengers. Selection is nothing like sharp enough to produce a slick machine. Still, the assumption that major features do in general have functions works reasonably well much of the time, and the assumption that they do not, or that they work for death, makes thought impossible (Mary Midgley, *Wickedness* [London: Routledge, 1984], 189).

28. Mary Midgley, *Evolution as a Religion: Strange Hopes and Stranger Fears* (London: Methuen, 1985), 6.

29. Malik, *Man, Beast and Zombie,* 191–92.

30. Frans DeWaal, the prominent primatologist, has also railed on this colossal confusion of categories. See Frans DeWaal, *Good Natured: The Origins of Right and Wrong in Humans and Other Animals* (Cambridge, MA: Harvard University Press, 1996), 16–17.

31. For example, Midgley, *Wickedness,* 189.

32. Midgley, *Beast and Man,* 142.

33. Mary Midgley, *Heart and Mind* (London: Methuan, 1983), 23.

34. For a focus on human universals as seen from cross-cultural anthropological comparisons, see Donald E. Brown, *Human Universals* (Philadelphia: Temple University Press, 1991).

35. Midgley, *Beast and Man,* 238. Her point, and explanation, is strongly reminiscent of Dewey: "Observation of a child, even a young baby, will convince the observer that a normal human being when awake is engaged in activity; he is a reservoir of energy that is continually overflowing. The organism moves, reaches, handles, pulls, pounds, tears, molds, crumples, looks, listens, etc." See Dewey and Tufts, *Ethics,* 343.

36. Midgley, *Beast and Man,* 279.

37. Ibid, 51–57.

38. Ibid, 56.

39. Ibid, 332.

40. "The more adaptable a creature is, the more directions it can go in. So it has more, not less, need for definite tastes to guide it. *What replaces closed instincts, therefore, is not just cleverness, but strong, innate, general desires and interests*" (Ibid, 332).

41. Midgley, *The Ethical Primate,* 136.

42. Midgley, *Wickedness,* 84; emphasis in quoted text appears in the original.

43. Midgley, *Science and Poetry*, 138.

44. Midgley, *Beast and Man*, 331.

45. "The general point is that other animals clearly lead a much more structured, less chaotic life than people have been accustomed to think, and are therefore, in certain definite ways, much less different from men than we have supposed. (There is still plenty of difference, but it is a different difference.)" (Midgley, *Beast and Man*, 25).

46. Ibid, 26.

47. Irenaus Eibl-Eibesfeldt, *Love and Hate* (New York: Holt, Rinehart, and Winston, 1972), 123–24, as quoted in Midgley, *Beast and Man*, 339.

48. Midgley, *Beast and Man*, 70.

49. On rape, see Randy Thornhill and Craig T. Palmer, *A Natural History of Rape*, (Cambridge, MA: MIT Press, 2000). See also Terry Burnham and Jay Phelan, *Mean Genes* (Cambridge, MA: Perseus Publishing, 2000); and Midgley, *Beast and Man*, 135.

50. Ibid, 182.

51. Ibid, 130.

52. Midgley, *The Ethical Primate*, 182.

53. Ibid, 180.

54. Midgley, *Animals and Why They Matter*, 106.

55. "Moreover, his method in the *Nicomachean Ethics* is exactly the one I am trying to follow here. He understands morality as the expression of natural human needs" (Midgley, *Beast and Man*, 260 footnote).

56. Midgley, *Heart and Mind*, 48.

57. A point made by Konrad Lorenz in "Introduction," in Charles Darwin, *The Expression of the Emotions in Man and Animals* (Chicago: University of Chicago Press, 1965, first published in 1872), xii.

58. Midgley, *Animals and Why They Matter*, 113; notice the similarity with Dewey's comment on wolves.

59. Ibid, 107.

60. "Here my point is the common-sense notion that our structure of instincts, as a whole, indicates the good and the bad for us. I am saying that, if Socrates is right in his facts, he is right in his argument. That is, if it is *true* that people are naturally inquiring animals, and if that inquiring tendency has a fairly central place in their natural structure of preference, that it follows that inquiry is an important good for them, that they ought not to stop each other from doing it (unless they have to), and should do it themselves, to an extent in proportion to the other things they also need to do. And so on for other tendencies" (Midgley, *Beast and Man*, 75).

61. "And if it makes any use of concepts like *dehumanization*, it presupposes a set of natural tendencies which will then be released to shape the human future. If people were really natureless, were mere indefinite lumps of dough moulded entirely by historical forces, we could have no notion at all of what they would be like or how they would feel in any other culture or epoch than our own" (Midgley, *Wickedness*, 106).

62. Midgley, *Evolution as a Religion*, 8; also, "people brought up in slave-owning communities have in many places gradually come to see that custom as wrong, and

have eventually abolished it. This was possible because they had within them other attitudes, other ideals, other perceptions, with which slave-owning conflicted. Those attitudes and ideals were an important part of their structure of feeling, even though they may not have been explicitly acknowledged in the culture" (Midgley, *The Ethical Primate*, 152–53).

63. Judith Hughes and Mary Midgley, *Women's Choices: Philosophical Problems Facing Feminism* (New York: St. Martins Press, 1983).

64. Midgley, *Heart and Mind*, 41.

65. Mary Midgley, *Can't We Make Moral Judgements?* (New York: St. Martins Press, 1993), 39.

66. Ibid, 39.

67. Ibid, 206–7.

68. Carl N. Degler, *In Search of Human Nature: The Decline and Revival of Darwinism in American Social Thought* (New York: Oxford University Press, 1991), 25–31.

69. For an in-depth discussion, see Degler, 105–38.

70. Midgley, *The Ethical Primate*, 8.

71. Ibid, 8.

72. Midgley, *Women's Choices*, 214.

73. Ibid, 189. As discussed in Chapter Two, Huxley made a similar point about childbearing and equality. It is not surprising that his 'feminism' would also see biology as, essentially, a burden.

74. Ibid, 172–73.

75. "Thus there was no sex discrimination. Pregnant women were excluded because they were pregnant, not because they were women. It was the disability and not the person or the group which was excluded. . . . Pregnant men would not be entitled to maternity benefits either" (Ibid, 161).

76. See Ibid, 165.

77. " . . . objectors might concede that people do have their own individual physical constitutions, and that the peculiarities of their brain and nervous system, and indeed of their whole physique, might make a great difference to their personality. (Were this not true, it would scarcely be possible to account for individuality at all.) But they would still reject the idea of a natural, characteristic difference in personality between the sexed. They view all character traits as in principle equally likely to be found in either sex. They see the distribution of them as entirely due to social forces" (Ibid, 207).

78. Ibid, 197.

79. In a fascinating case, recently published, a boy who suffered a botched circumcision had surgery to alter his gender, and was raised as a girl. The book documents the child's inability to live as a girl, in spite of the fact that he never knew that he had been born as a boy, and his choice in adulthood, once learning of his history, to live as a male. See John Colapinto, *As Nature Made Him: The Boy Who Was Raised As a Girl* (New York: HarperCollins Publishers, 2001).

80. Midgley, *Women's Choices*, 199.

81. "Is the difference between men and women natural, or is it produced by culture? Here is another false antithesis. In other fields today this one is usually nailed as false quite quickly, being unkindly referred to as 'the old nature–nurture

controversy'. . . . The two sets do not compete. The job of working out their details and relative importance is done by empirical enquiry, not by dying on the barricades" (Ibid, 185).

82. The first and most famous description being Carol Gilligan's *In a Different Voice* (Cambridge, MA: Harvard University Press, 1993), which leaves the question of biological versus sociological origins open. 83. "Now in theory this skill [argument] is always supposed to be linked with the deeper and more serious one of following an argument disinterestedly wherever it may lead, as a path to the truth. But in practice the aims of truth and persuasion diverge so sharply that an obsessive interest in winning arguments must eventually kill the passion for truth. They cannot survive together" (Ibid, 202–3). See also Ibid, 206.

84. Ibid, 206.

85. "Their [Sartre and Nietzsche] peculiar idea of human dignity and independence is specifically a male one and is designed as such. One point of it is to elevate the male condition and represent the female one as degraded. It is just as easy, if one is interested in this sort of game, to exalt the female as being the only one who can break through the bounds of solitude, who can have the mystical experience of being both one and two, and who therefore is not afraid of otherness and constitutes our species's link with the glories of the physical universe. Either sex can, if it likes, claim superiority. You can take your choice. But it is best to take it while remembering that both parties are here for keeps" (Ibid, 215).

86. Ibid, 222.

87. "The depressing fact that extreme individualism is exploitative has been blurred for us by more than a century of romantic propaganda which pilloried the bourgeois vices, ignored the bourgeois virtues, and sentimentally indulged the individualistic vices—conceit, ingratitude, self-dramatization, infidelity, hardheartedness, uselessness and the fear of intimacy. This is a morality tailored to fit only adolescents at the moment of leaving home, young Werthers who remain perpetually young. Women's life-cycle is such that, even if they enter this narcissistic realm, they can seldom stay in it" (Ibid, 222–23).

88. Ibid, 222.

89. "So far we have argued that men and women are biologically different and that the difference is and ought to be important—even in legislation—if women are to be treated fairly. We have also said that equal treatment under the law sometimes acts to the disadvantage for some women. What are we saying? That we do not always want equal treatment. This is beginning to sound subversive. Are we then saying that what we want is unequal treatment and unequal rights?" (Ibid, 171).

90. David Hume, *A Treatise of Human Nature* (Oxford: Clarendon, 1964).

91. Midgley, *Wickedness*, 200.

92. Midgley, *Beast and Man*, 282. Also, 191.

93. "Morality, then, is our way of dealing with the up-and-down dimension which everybody who thinks seriously about human life must see as our central problem. What makes this dealing so hard, however, is the constant ambivalence, the way in which nearly every feature of human life can be described and thought of either more or less favourably" (Midgley, *Wickedness*, 194); also Midgley, *Can't We Make Moral Judgements?*, 164.

94. "Atomism is not an option. We could not start to discriminate between anger and ambition, habit and jealousy as possible motives for an act unless we had an idea of the framework within which they work, of the kind of total character to which they must all belong. However obscure the idea of the whole may be in detail, its general shape is essential for explanation. And we do have that idea. No doubt if we ourselves did not also exemplify it—if we were members of an alien species with a quite different pattern of motives—we would have found it very hard to construct such a scheme. But this is a piece of bad luck which we do not have and need not imitate. We approach the problems of human psychology as humans, and it seems a pity to waste that advantage" (Midgley, *Heart and Mind*, 166).

95. "Serious neglect of cubs, or brutal treatment of them, would be thoroughly unnatural among wolves. So would disrespectful or uncooperative behavior to elders. These things sometimes happen. But they are not just unfortunate, they are out of character; they show something wrong, something, as Butler said, 'disproportionate to their nature as a whole.'" (Midgley, *Beast and Man*, 279).

96. "Man is not held together only by consistent thought or by the shape of the surrounding landscape. He holds himself together (up to a point and with great difficulty) by virtue of a *feeling*—a strong wish for order and unity" (Midgley, *Heart and Mind*, 86). Also: "And the integration of the personality is—on Darwin's suggestion as on Jung's—a primary need, without which nothing else is possible" (Midgley, *Wickedness*, 185).

97. Midgley, *Beast and Man*, 80.

98. Ibid, 80.

99. On being raw material from which human beings choose good or evil, see Ibid, 70. On natural desires which can destroy the system, and are therefore, at least in a weak sense, bads, see Ibid, 328.

100. Ibid, 182.

101. Midgley, *The Ethical Primate*, 80–91.

102. Midgley, *Wickedness*, 79.

103. Ibid, 79.

104. Ibid, 80.

105. For example, Midgley, *Beast and Man*, 41.

106. Ibid, 169.

107. Midgley believes this to be Darwin's argument: "In searching, then, for the special force possessed by 'the imperious word *ought*,' he [Darwin] pointed to the clash between the chronic social affections and the acute but transient motives which often oppose them. Intelligent beings would, he concluded, naturally try to produce rules which would protect the priority of the first group" (Midgley, *The Ethical Primate*, 140).

108. Ibid, 171–72.

109. Ibid, 173.

110. Midgley, *Beast and Man*, 41.

111. "[Darwin] is insisting that no creature with inner conflicts of this gravity can avoid taking sides somehow. *This* is what makes morality necessary. There is no semi-paradisal option of simply letting things take their course. Once you realize that you are constantly wrecking your own schemes in the way that the migrating

swallow does, you are forced to evolve some sort of priority system and to try to stick to it. That means having a morality. If, therefore, 'immoralism' is taken to mean having no morality, then it is not a possible option" (Midgley, *The Ethical Primate*, 179).

112. Ibid, 181.
113. Midgley, *Beast and Man*, 40.
114. Midgley, *The Ethical Primate*, 148.
115. "We tend to think of animals as not having this problem. *They do have the problem. What they do not have is our way of solving it by thinking about it. But they still have a way of solving it-* namely, by a structure of motives that shapes their lives around a certain preferred kind of solution. If we did not have that too, thinking would get us nowhere" (Midgley, *Beast and Man*, 280).
116. Ibid, 162–63.
117. Midgley, *The Ethical Primate*, 149.
118. "Strong feeling is fully appropriate to well grounded belief on important subjects. Its absence would be a fault. This is the element of truth in Emotivism; morality does require feeling. The emotivist's mistake is in supposing that it requires nothing else; in trying to detach such feelings from the thoughts that properly belong to them" (Midgley, *Animals and Why They Matter*, 35).
119. Midgley, *Can't We Make Moral Judgements?*, 155.
120. Midgley, *Heart and Mind*, 102; John Dewey, *Human Nature and Conduct*, 33.
121. Midgley, *Can't We Make Moral Judgements?*, 100.
122. Dewey, *Human Nature and Conduct* (New York: Henry Holt, 1922), 25.
123. See, for example, Midgley's citing of C. S. Lewis's explanation of prayer as an embodied activity (in Midgley, *Beast and Man*, 310). Also, her story about Bertrand Russell's son, who, in spite of Russell being convinced that he had rationally explained to his son why one should not be afraid of the dark, his son remained petrified of the dark for years after (see Midgley, *Women's Choices*, 203).
124. "Balance, in fact, is not just a negative matter of not falling over; it is a positive one of attaining one's full growth" (Midgley, *Beast and Man*, 192). While Dewey and Midgley are quite similar in their search for balance, Midgley is explicitly Aristotelian in that she argues that there are discrete ends to a human life, whereas Dewey maintains that growth is the only ultimately ends to a human life.
125. Midgley, *The Ethical Primate*, 23.
126. Midgley, *Beast and Man*, 193, 280; and Midgley, *Heart and Mind*, 1.
127. For example, in Robert Ardrey, *The Territorial Imperative* (London: Collins, 1967), 288–305, 332–344. Also, Robert Wrangham and Dale Peterson, *Demonic Males: Apes and the Origins of Human Violence* (Boston: Houghton Mifflin, 1996); and Michael Ghiglieri, *The Dark Side of Man: Tracing the Origins of Male Violence* (New York: Perseus Books, 1999).
128. Midgley, *Can't We Make Moral Judgements?*, 33.
129. Midgley, *Wickedness*, 49.
130. Midgley, *Beast and Man*, xxviii.
131. Midgley, *Wickedness*, 59–60.
132. "The positive motives which move them may not be bad at all; they are often quite decent ones like prudence, loyalty, self-fulfillment and professional

conscientiousness. The appalling element lies in the lack of the *other* motives which ought to balance these—in particular, of a proper regard for other people and of a proper priority system which would enforce it" (Midgley, Wickedness, 67).

133. Ibid, 155.
134. Midgley, *Wickedness*, 59.
135. Ibid, 55.
136. Ibid, 67.
137. Ibid, 117–123.
138. "For instance, it is largely in this sort of unthinking way that we in the more prosperous countries of the world are now engaged in starving out the poorer ones" (Ibid, 74–75).
139. Ibid, 63.
140. Hannah Arendt, *Eichmann in Jerusalem: A Report on the Banality of Evil* (New York: Penguin Books, 1976, first published 1963).
141. Midgley, *Wickedness*, 64.
142. As quoted in Mary Midgley, *Wisdom, Information and Wonder* (London: Routledge, 1989), 156.
143. David Hume, *Treatise of Human Nature*, as quoted in Michael Ruse, *Taking Darwin Seriously: A Naturalistic Approach to Philosophy* (Oxford: Basil Blackwell, 1987), 87.
144. The example of sociability in horses was taken from Janet Radcliffe Richards, *Human Nature after Darwin: A Philosophical Introduction* (London: Routledge, 2000), 230.
145. Midgley, *Wisdom, Information, and Wonder*, 158.
146. Midgley, *Beast and Man*, xliii.
147. Ibid, 75.
148. "Putting the point another way, we all believe that understanding what we are naturally fit for, capable of, and adapted to will help us to know what is good for us and therefore, to know what to do. Common sense takes this view; so does Freudian psychology. So does traditional moral philosophy" (Ibid, 177).
149. Midgley, *Wisdom, Information and Wonder*, 111.
150. Ibid, 18.
151. Ibid, 155.
152. Midgley, *Evolution as a Religion*, 135.
153. Midgley, *Beast and Man*, 156.
154. Ibid, 140.
155. Midgley, *Beast and Man*, 18; for a similar sentiment, see also Midgley, *Wisdom, Information and Wonder*, 6.
156. Midgley's *Science and Poetry* (2000), published after Richard Dawkins's *Unweaving the Rainbow: Science, Delusion, and the Appetite for Wonder* (London: Penguin, 1998), demonstrates the difference between the two as they relate to science and the humanities. Midgley wants scientists to have a better understanding of the larger conceptual background within which they work. Dawkins wants poets to understand better the facts of the world so that their work can be more relevant.
157. Midgley, *Wisdom, Information and Wonder*, 9.
158. Midgley, *Science and Poetry*, 70.

159. Ibid, 147.
160. See Ibid, 192. Also Midgley, *Can't We Make Moral Judgements?*, 60.
161. Midgley, *Beast and Man*, 288; also: "Freedom itself is a negative ideal. Its meaning depends in each case on what particular bonds it frees us *from*. The reformers who fought each special kind of oppression were always led by a vision of a particular kind of freedom that would replace it, a special way in which society would be changed when they had cut a certain kind of bond. But it has gradually become plain that this bond-cutting sequence is cumulative, which means that it cannot go on for ever. Humans are bond-forming animals. When all the bonds are cut—when the various kinds of freedom are all added together—when a general vision of abstract freedom from every commitment replaces the more limited aims—then, it seems, we might be left with a meaningless life. It begins to seem doubtful whether any kind of human society is then possible at all" (Midgley, *Science and Poetry*, 14).
162. DeTocqueville, as quoted by Midgley: "I have shown how it is that, in ages of equality, every man seeks for his opinions within himself: I am now to show how it is that, in these same ages, all his feelings are turned towards himself alone. Individualism is a novel expression, to which a novel idea has given birth. Our fathers were only acquainted with egoisme (selfishness). Selfishness is a passionate and exaggerated love of self, which leads a man to connect everything with himself, and to prefer himself to everything in the world. Individualism is a mature and calm feeling, which disposes each member of the community to sever himself from the mass of his fellows, and to draw apart with his family and his friends: so that, after he has thus formed a little circle of his own, he willingly leaves society at large to itself. Selfishness originates in blind instinct: individualism proceeds from erroneous judgment more than from depraved feelings: it originates as much in deficiencies of mind as in perversity of heart" (Ibid, 152).
163. "This means acknowledging our kinship with the rest of the biosphere. If we do not feel perfectly at home here, that may after all have something to do with the way in which we have treated the place. Any home can be made uninhabitable.... Our dignity arises *within* nature, not against it" (Midgley, *Beast and Man*, 196).
164. Midgley, *Science and Poetry*, 14.
165. "Obviously the idea of Gaia is a myth, a symbol. But then so is the sociobiological idea of the Selfish Gene. One of these myths emphasizes our separateness from the world around us. The other emphasises our profound dependence on it. Since wholes are quite as real as parts, there is no reason in principle why we should have to prefer the first emphasis over the second. The choice between them depends on their relevance to our situation. And given that current situation, there seems to me to be little doubt about which of them we most need to guide our thinking today" (Ibid, 17).
166. Midgley, *Beast and Man*, 19.
167. Midgley, *Evolution as a Religion*, 112.
168. Midgley, *Beast and Man*, 362.
169. Midgley, *Animals and Why They Matter*, 145.
170. Ibid, 145.
171. Midgley, *Science and Poetry*, 184.
172. Midgley, *Evolution as a Religion*, 160.

CHAPTER FIVE. A DARWINIAN EDUCATION

1. Neil Postman, *The End of Education: Redefining the Value of School* (New York: Vintage Books, 1996), 19.

2. Ibid, 5.

3. Alven Neiman, "Pragmatism: The Aims of Education and the Meaning of Life," in Wendy Kohli, ed., *Critical Conversations in Philosophy of Education* (London: Routledge, 1995), 56–72.

4. See Alistair MacIntyre, *After Virtue* (Notre Dame, IN: University of Notre Dame Press, 1981), 56, 81, 83.

5. "In that [Aristotelian/classical] context moral judgments were at once hypothetical and categorical in form. They were hypothetical in so far as they expressed a judgment as to what conduct would be teleologically appropriate for a human being: 'You ought to do so-and-so, if and since your *telos* is such and such' or perhaps 'You ought to do so-and-so, if you do not want your essential desires to be frustrated'" (Ibid, 57). Also, "Every activity, every enquiry, every practice aims at some good; for by 'the good' or 'a good' we meant that at which human beings characteristically aim. It is important that Aristotle's initial arguments in the *Ethics* presuppose that what G. E. Moore was to call the 'naturalistic fallacy' is not a fallacy at all and that statements about what is good—and what is just or courageous or excellent in other ways—just are a species of factual statement. Human beings, like the members of all other species, have a specific nature; and that nature is such that they have certain aims and goals, such that they move by nature towards a specific *telos*. The good is defined in terms of their specific characteristics. Hence Aristotle's ethics, expounded as he expounds it, presupposes his metaphysical biology" (Ibid, 139).

6. "A key part of my thesis has been that modern moral utterance and practice can only be understood as a series of fragmented survivals from an older past and that the insoluble problems which they have generated for modern moral theorists will remain insoluble until this is well understood. If the deontological character of moral judgments is the ghost of conceptions of divine law which are quite alien to the metaphysics of modernity and if the teleological character is similarly the ghost of conceptions of human nature and activity which are equally not at home in the modern world, we should expect the problems of understanding and of assigning an intelligible status to moral judgments both continually to arise and as continually to prove inhospitable to philosophical solutions" (Ibid, 105).

7. ". . . any adequate generally Aristotelian account must supply a teleological account which can replace Aristotle's metaphysical biology" (Ibid, 152). Also, "How far, for example, and in what ways is it Aristotelian? It is—happily—not Aristotleian in two ways in which a good deal of the rest of the tradition also dissents from Aristotle. First, although this account of the virtues is teleological, it does not require the identification of any teleology in nature, and hence it does not require any allegiance to Aristotle's metaphysical biology. And secondly, just because of the multiplicity of human practices and the consequent multiplicity of goods in the pursuit of which the virtues may be exercised—goods which will often be contingently incompatible and which will therefore make rival claims upon our

allegiance—conflict will not spring solely from flaws in individual character. But it was just on these two matters that Aristotle's account of the virtues seemed most vulnerable; hence if it turns out to be the case that this socially teleological account can support Aristotle's general account of the virtues as well as does his own biologically teleological account, these differences from Aristotle himself may well be regarded as strengthening rather than weakening the case for a generally Aristotelian standpoint" (Ibid, 183).

8. However, in MacIntyre's Carus Lectures in 1997, MacIntyre seems to have shifted his position, and begun to take biology seriously from within an Aristotelian framework. See Alistair MacIntyre, *Dependent Rational Animals* (Chicago, IL: Open Court, 1999).

9. See Chapter Two on Dewey, 101–5, and John Dewey, "The Influence of Darwinism on Philosophy," in Larry Hickman and Thomas Alexander, eds., *The Essential Dewey: Volume I* (Bloomington and Indianapolis: Indiana University Press, 1998), 40–41.

10. Ibid, 40–41.

11. MacIntyre, *After Virtue*, 53.

12. Mary Midgley, *Beast and Man* (London: Routledge, 1995 [1978]), 41.

13. See Midgley, *Science as Salvation* (London: Routledge, 1992), 165–224.

14. Ibid, 183–94.

15. MacIntyre makes a similar point: "It was in the seventeenth and eighteenth centuries that morality came generally to be understood as offering a solution to the problems posed by human egoism and that the content of morality came to be largely equated with altruism. For it was in that same period that men came to be thought of as in some dangerous measure egoistic by nature; and it is only once we think of mankind as by nature dangerously egoistic that altruism becomes at once socially necessary and yet apparently impossible and, if and when it occurs, inexplicable. On the traditional Aristotelian view such problems do not arise. For what education in the virtues teaches me is that my good as a man is one and the same as the good of those others with whom I am bound up in human community. There is no way of my pursuing my good which is necessarily antagonistic to you pursuing yours, because the good is neither mine peculiarly nor yours peculiarly—goods are not private property. Hence Aristotle's definition of friendship, the fundamental form of human relationship, is in terms of shared goods. The egoist is thus, in the ancient and medieval world, always someone who has made a fundamental mistake about where his own good lies and someone who has thus and to that extent excluded himself from human relationships" (Ibid, 213).

16. See, for example, John Cottingham, "Neo-Naturalism and its Pitfalls," *Philosophy* 58 (1983): 465; Stephen R. L. Clark, "The Absence of a Gap Between Facts and Values," *Aristotelian Society Supplementary Volume* 54 (1980); Clifford Geertz, "Anti Anti-Relativism," in Geertz, *Available Light: Anthropological Reflections on Philosophical Topics* (Princeton: Princeton University Press, 2000), 51–59.

17. Thomas Eagleton, *The Ideology of the Aesthetic* (Oxford: Basil Blackwell, 1990), 235, as quoted in Marjorie O'Loughlin, "Reinstating Emotion in Educational Thinking," in *Philosophy of Education 1997* (Normal, IL: Philosophy of Education Society, 1998), 409.

18. John Dewey, *Human Nature and Conduct,* Volume 14 in Jo Anne Boylston, ed., *John Dewey: The Middle Works* (Carbondale: University of Southern Illinois Press, 1976), 4.

19. Midgley, *Beast and Man,* 52–53.

20. Ibid, 55.

21. Paul Hirst, "Educational Aims: Their Nature and Content," *Philosophy of Education 1991* (Normal, IL: Philosophy of Education Society, 1992), 40–53.

22. "The justification for that notion of the good life was taken to lie, first, in certain forms of transcendental argument which held that there can be no more ultimately justifiable pursuits than the intrinsically worthwhile pursuits of reason in all its forms and, second, in the successful ordering of all other human concerns in terms made possible by the achievement of reason into a coherent and consistent whole" (Ibid, 41–42).

23. Israel Scheffler, "In Praise of the Cognitive Emotions," in Scheffler, *In Praise of the Cognitive Emotions and Other Essays in the Philosophy of Education* (New York and London: Routledge, 1991), 3–17.

24. Richard S. Peters, "The Education of the Emotions," in Dearden, Hirst and Peters, eds., *Education and the Development of Reason* (London and Boston: Routledge and Kegan Paul, 1972), 466–83.

25. As Hirst described it: "In keeping with this emphasis was the view that the central function of cognitive capacities is the formation of conceptual schemes in which judgments of truth can be made, and that thence can be achieved bodies of justifiable or rational beliefs, rational actions, and indeed rational emotions" (Hirst, 41).

26. "I had been advised early in life that sound decisions came from a cool head, that emotions and reason did not mix any more than oil and water. I had grown up accustomed to thinking that the mechanisms of reason existed in a separate province of the mind, where emotion should not be allowed to intrude, and when I thought of the brain behind that mind, I envisioned separate neural systems for reason and emotion" (Damasio, xi). Peters and Scheffler do not proscribe to this view, but rather that our minds should evaluate the contribution of the emotions to cognitive development, and train the emotions accordingly.

27. See, for example, Marjorie O'Loughlin, "Reinstating Emotion in Educational Thinking," *Philosophy of Education 1997* (Normal, IL: Philosophy of Education Society, 1998), 404.

28. Ibid, 404.

29. Clive Beck and Clare Madott Kosnik, "Caring for the Emotions: Toward a More Balanced Schooling," *Philosophy of Education 1995,* (Normal, IL: Philosophy of Education Society, 1999), 161.

30. "They think this can be done by transferring human minds to computers so as to prolong their existence into an epoch when there will be no other organized matter at all—a time when there will be nothing to do except (presumably) to consider and communicate abstractions. In the absence of anything to talk about, it is not even clear what the topic of conversations will be, except perhaps mathematics. . . . It is the further assumption about values, the assumption that the life which they would then live—a life without sense-perception or emotion or the power to act, a life consisting solely in the arrangement of abstract 'information'—would be

a human life, or indeed anything that could intelligibly be called life at all" (Mary Midgley, *The Ethical Primate* [London: Routledge, 1994], 10).

31. Aristotle, *Ethics*, 328–35.

32. There is much empirical evidence which supports Midgley's philosophical position. See, for example, Antonio Damasio, *Descartes' Error* (New York: Avon Books, 1994), 178.

33. See a discussion on the emergence of consciousness as an adaptation to maximize the evolutionary advantages of emotions in Antonio Damasio, *The Feeling of What Happens* (San Diego: Harcourt, 1999), especially 284–85.

34. Hirst, 44.

35. See, for example, Mary Midgley, *Beast and Man*, 130.

36. "It is not surprising, then, that alongside the 'rationalist' approach previously outlined, an alternative account of educational aims developed that was based on more 'utilitarian' presuppositions. In this the capacities of reason were seen as necessarily exercised in the service of substantive, naturally given wants and desires, whose satisfaction is, in certain cases, necessary for human survival and, in all cases, fundamental to human well-being. From this point of view, the conceptual schemes by which we make sense of our environment, ourselves, and our actions are primarily concerned with the achievement of satisfactions we seek in response to natural functions and in the exercise of our naturally given capacities. The exercise of reason and the knowledge and understanding it makes possible are primarily practical in their significance. They are concerned with the taking of means to ends which are in the last analysis naturally given. And in taking those means, reason helps us to order our wants, achieve maximum satisfaction, and provide priorities when wants conflict" (Hirst, 44).

37. Ibid, 46.

38. *Beast and Man*, 130.

39. Ibid, 280.

40. Damasio gives scientific support for such a claim as to the embodied nature of our reason from a neurological perspective. Damasio shows how reason structurally emerges from the emotional structures of the brain, so that the two processes are intimately connected with one another. Individuals with damage to the supposedly emotional structures of the brain, with accompanying loss of human emotions, exhibit damage to the functioning of proper reasoning, as well. Our rationality is, in fact, dependent on our emotions to give it structure and purpose. Without it, reason can only distinguish a loose collection of facts with no context or purpose within which to give it structure and meaning. Damasio argues that our feelings (the conscious experience of emotions) allow us to avoid trial-and-error reasoning, giving us clues as to what is best for the individual, and therefore shortcuts for evaluating our proper response. 'I feel scared' means that my emotions are telling me that I have good reason to feel scared, and that is the starting point of evaluating what to do. Emotions can avoid conscious reflection altogether, allowing us prerational and instinctive response to events without any conscious decision making. Feelings, as Damasio asserts, should be taken seriously. They have evolved to tell us something about ourselves and the world. See Damasio, *Descartes'*

Error, 9, 53–58, 128,171–72, 189, 200. Also in Daniel Goleman, *Emotional Intelligence* (New York: Bantam Books, 1995), 9.

41. A phrase first coined by Gilbert Ryle, *The Concept of Mind* (London: Penguin Books, 1949).

42. On Aristotle, see *Ethics,* 91–92.

43. Aristotle, 91–92; Mary Midgley, *Can't We Make Moral Judgements?* (New York: St. Martins Press, 1993), 100, 155.

44. *Beast and Man,* 309–10, 313.

45. "We repudiated all versions of the doctrine of Original Sin, of there being insane and irrational springs of wickedness in most men. We were not aware that civilization was a thin and precarious crust. . . . We had no respect for traditional wisdom or the restraints of custom. We lacked reverence, as Lawrence observed and as Ludwig [Wittgenstein] also used to say, for everything and everyone. It did not occur to us to respect the extraordinary accomplishment of our predecessors in the ordering of life (as it now seems to me to have been) or the elaborate framework which they had devised to protect this order" (as quoted in Mary Midgley, *Heart and Mind* [London: Methuan, 1983], 64).

46. See Shlomo Avineri and Avner de-Shalit, "Introduction," in Avineri and de-Shalit, eds., *Communitarianism and Individualism* (Oxford: Oxford University Press, 1992), 10.

47. Martha Nussbaum, "Patriotism and Cosmopolitanism," in Joshua Cohen, ed., *For Love of Country* (Boston: Beacon Press, 1996), 5–17.

48. Ibid, 5–17.

49. For example: "If one begins life as a child who loves and trusts his or her parents, it is tempting to want to reconstruct citizenship along the same lines, finding in an idealized image of a nation a surrogate parent who will do one's thinking for one. Cosmopolitanism offers no such refuge; it offers only reason and the love of humanity, which may seem at times less colorful than other sources of belonging" (Ibid, 15). But Nussbaum's position is, already in *For Love of Country,* perhaps more nuanced than I am suggesting, and its more embodied dialogue between reason and emotions is fully explored in her book, Martha Nussbaum, *Upheavals of Thought: The Intelligence of Emotions* (Cambridge: Cambridge University Press, 2001).

50. Midgley's Darwinist leftist agenda stands in stark contrast to the philosopher Peter Singer's articulation of a Darwinian agenda for the left. His version of a progressive Darwinism extends from a combination of his utilitarian philosophy applied to the Huxley–Dawkins interpretation of Darwinism. See Peter Singer, *A Darwinian Left: Politics, Evolution and Cooperation* (New Haven and London: Yale University Press, 2000).

51. Evaluating Midgley's hermeneutics through Gallagher's typology, I believe that Midgley, and Dewey as well, can be associated most closely with Gadamar's approach, which Gallagher defines as a "moderate hermeneutics," as opposed to a conservative, critical or deconstructionist one. See Shaun Gallagher, *Hermeneutics and Education* (Albany: State University of New York Press, 1992), 9–11.

52. Elliott Sober and David Sloan Wilson, *Unto Others: The Evolution and Psychology of Unselfish Behavior* (Cambridge, MA: Harvard University Press, 1998), 9. It's a point already made by Darwin.

53. Mary Midgley, *Animals and Why They Matter* (Athens, GA: University of Georgia Press, 1984), 109.

54. Ibid, 108.

55. This is because, for Midgley, our connection with other human beings is prerational, and our division of human beings into "us" and "them" is cultural. Nussbaum, in her response to critics of her "Patriotism and Cosmpolitanism," makes a similar point, which seems in dissonance with her earlier portrayal of reason as the avenue from which to access our common humanity: "At birth, all an infant is is a human being. Its needs are the universal needs for food and comfort and light. Infants respond, innately, to the sight of a human face. A smile from a human being elicits a reactive smile, and there is reason to think this an innate capacity of recognition. At the same time, in the first few months of life an infant is also getting close experience of one or more particular people, whom it soon learns to tell apart from others, roughly at the time that it is also learning to demarcate itself from them. These people have a culture, so all the child's interactions with them are mediated by cultural specificity; but they are also mediated by needs that are in some form common, and that form the basis for later recognition of the common" Nussbaum, "Reply," in Cohen, ed., 142.

56. Midgley, *Animals and Why They Matter*, 118.

57. Nussbaum suggests a similar argument. See Nussbaum, "Reply," 144.

58. "I cannot pursue this fascinating point here further than to say that I prefer to formulate it, as Eric Berne does, by saying that the Child still survives within each of us; that it is not a passing role, but a lasting aspect of our character. "Actually" Berne writes, "the Child is in many ways the most valuable part of the personality, and can contribute to the individual's life exactly what an actual child can contribute to family life: charm, pleasure and creativity." Midgley, *Beast and Man*, 342.

59. Avital and Jablonka document how widespread culturally learned behavior is in the natural world. Eitan Avital and Eva Jablonka, *Animal Traditions: Behavioural Inheritance in Evolution* (Cambridge: Cambridge University Press, 2000), 1.

60. Midgley, *Women's Choices* (New York: St. Martins Press, 1983), 210–11.

61. Ibid, 210–11.

62. Midgley quotes Iris Murdoch, friend and mentor, who articulates a similar sentiment: "Words are the most subtle symbols that we possess and our human fabric depend on them. . . . It is totally misleading to speak, for instance, of 'two cultures', one literary–humane and the other scientific, as if these were of equal status. There is only one culture, of which science, so interesting and so dangerous, is now an important part. But the most essential and fundamental aspect of culture is the study of literature, since this is an education in how to picture and understand human situations. We are men and we are moral agents before we are scientists, and the place of science in human life must be discussed in *words*. That is why it is and always will be more important to know about Shakespeare than to know about any scientist, and if there is a Shakespeare of science his name is Aristotle" (Iris Murdoch, *The Sovereignty of the Good* [London and New York, 1996], 34), as appearing in Midgley, *Utopias, Dolphins and Computers*, 60.

63. Mary Midgley, *Evolution as a Religion: Strange Hopes and Stranger Fears* (London: Methuen, 1985), 4.

64. Mary Midgley, *Wickedness* (London: Routledge, 1984), 84.

65. See discussion in Frans De Waal, *The Ape and the Sushi Master* (New York: Basic Books, 2001), 149–76.

66. See Midgley, *Wickedness*, 84.

67. Midgley, *Utopias, Dolphins and Computers*, 62–63.

68. Ibid, 65.

69. Ibid, 63.

70. Midgley, *Animals and Why They Matter*, 107. Similarly, "We need the natural, sincere reactions of those around us if we are to locate ourselves morally or socially at all. They give us our bearings in the world. No child ever grows up without constantly experiencing both disapproval and approval, and the serious possibility that both will continue is essential for our lives. Sometimes we need to accept disapproval and to learn from it, sometimes to soften it by friendliness and argument, sometimes to persist in spite of it. But, if we did not know that it was there or understand its grounds, we could not begin to do any of these things" (Mary Midgley, *Wisdom, Information and Wonder* [London: Routledge, 1989], 170). Also, see Mary Midgley, *Science and Poetry* (New York and London: Routledge, 2001), 90–91.

71. Midgley, *Science and Poetry*, 51. Midgley ties Mill's insight to Mill's critique of his father's moral relation to him and his siblings: "The natural consequence was that educators in the Age of Reason not only typically ignored the development of the feelings but often tried, so far as possible, to suppress them entirely. Among the many autobiographers who describe the effects of that system on their lives, John Stuart Mill put the point well. His father (he writes) was not unkind, but he 'never varied in rating intellectual enjoyments above all others. . . . For passionate emotions of all sorts, and for everything which has been said or written in exaltation of them, he professed the greatest contempt. . . . The element which was chiefly deficient in his moral relation to his children was that of tenderness" (Ibid, 51).

72. See Jane Roland Martin, *The Schoolhome* (Cambridge, MA: Harvard University Press, 1992); Jane Roland Martin, "Education for Domestic Tranquility," in Wendy Kohli, ed., *Critical Conversations in Philosophy of Education* (New York and London: Routledge, 1995), 45–55; and Nel Noddings, *The Challenge to Care in Schools* (New York: Teachers College Press, 1992).

73. Martin, 122, quoted in Beck and Kosnick, 166.

74. Beck and Kosnik, 166–68.

75. "A rich emotionality is essential to well-being—and even to academic learning—for both students and teachers. The time has come for schools to give up the artificial and damaging separation between emotion and cognition. As we attempt to engage in "emotional education," however, it would be a mistake to try to do so by means of academic courses and units of study which reinforce the idea that cognition is the gateway to the good life. We need to create classrooms and schools which are genuine communities, within which students and teachers quite naturally experience the joys and learn the skills of emotional living" (Beck and Kosnik, 168).

76. Martin, 85, quoted in Beck and Kosnick, 165.

77. See, for example, Michael Walzer's response to Martha Nussbaum in Joshua Cohen, ed., 126.
78. Midgley, *Animals and Why They Matter*, 108.
79. Ibid, 117.
80. Ibid, 109.
81. Nussbaum, "Introduction," in Cohen, ed., 11–12.
82. Midgley, *Animals and Why They Matter*, 106.
83. Midgley quotes Maurice Wilkins in this context: "The poet Coleridge is said to have claimed that a scientist must love the object he studies, otherwise he could not respond to its true nature. I believe Coleridge's idea of love expresses the ideal scientific attitude as well as or better than the ideal of curiosity, which has been part of the scientific tradition of objective, value-free enquiry. Love includes curiosity, but curiosity need not include love. If we eulogize curiosity, we run the risk of encouraging a scientist to be like a child who, in its intense desire to know, tears a butterfly to pieces" (Maurice Wilkins, "The Nobility of the Scientific Enterprise," *Interdisciplinary Science Reviews* 10:1 (1985), as quoted in Midgley, *Wisdom, Information and Wonder*, 39–40.
84. Midgley, *Wisdom, Information and Wonder*, 40.
85. Midgley, *Science and Poetry*, 185.
86. Midgley, *Wisdom, Information and Wonder*, 41.
87. Midgley, *Beast and Man*, 362
88. Ibid, 362.
89. Midgley, *Wisdom, Information and Wonder*, 39.
90. After a lengthy quote from the Book of Job, Midgley concludes her book *Beast and Man* as follows: "That is the sort of way Charles Darwin looked at the physical universe, and, unless I am much mistaken, Aristotle too. That is the sort of universe in which our nature is adapted to live, not one alien and contemptible to us, from which we must be segregated. As I understand Humanism, this is its message. Humanism cannot only mean destroying God; its chief job is to understand and save man. But man can neither be understood nor saved alone" (Midgley, *Beast and Man*, 363).
91. "When some portion of the biosphere is rather unpopular with the human race—a crocodile, a dandelion, a stony valley, a snowstorm, and odd-shaped flint—there are three sorts of human being who are particularly likely still to see point in it and befriend it. They are poets, scientists and children. Inside each of us, I suggest, representatives of all these groups may be found. The decision whether to go in for minimalism is the decision whether to suppress them or to take their advice" (Midgley, *Animals and Why They Matter*, 145).
92. Postman, 4–18.
93. Ibid, 19.
94. Alan Ryan, *John Dewey and the High Tide of American Liberalism* (New York: W. W. Norton and Company, 1995), 130.

Index

acculturation process, 35–36, 129, 147, 148–153
adaptation
    passing to next generation, 94
    in sociobiology, 90
aesthetics, science versus, 39–40
After Virtue (MacIntyre), 130–131
aggression
    as fundamental need, 96–97, 115, 156
    Lorenz on, 96–97
    love and, 96
    motivation for, 107
    sympathy versus, 95
Alexander, Thomas M., 174*n*.20
alienation, separation from natural world and, 133
altruism
    emergence of, 9–10
    of morality, 32
    selfishness versus, 13
    Wilson on, 90
animal behavior
    children and, 160
    conflicting desires, 12, 109–110
    ethology and, 95
    human nature versus, 19–22, 136, 139
    instincts in, 93–94, 109–110, 139
    maternal instinct and, 12, 94, 102–105, 109–110
    motivation in, 87, 95–96
    needs in, 98
    play, 93, 160
    social animal natures, 11

animal rights movement, 126
anthropocentrism, 162
anthropology
    decline of Darwinism and, 85–86
    human nature and, 92–93
Arendt, Hannah, 117–118
Aristotelian worldview
    Darwinian biology and, 4, 32, 130–132
    Dewey and, 60, 61, 78, 130–132, 145–146, 152
    MacIntyre and, 130–132, 199*n*.15
    Midgley and, 4, 91–92, 97–98, 108–109, 116–117, 142, 145–146, 152, 163, 203*n*.62
Aristotle, 60
    human nature and, 12–13, 108–109, 134, 139–140
    Job and, 163
    on morality, 113, 115, 145–147, 198–199*n*.7, 199*n*.15
    tree metaphor of, 114
    on wrongdoing, 116
art, 155
artificial selection, 18, 172*n*.64
atomistic view, 124–128
authenticity, Spencer and, 39

banality of evil (Arendt), 117–118
Barbour, Ian, 172*n*.70
Beast and Man (Midgley), 86, 189*n*.12, 189*n*.24, 190*n*.35, 191*n*.45, 191*n*.60, 194*n*.95, 195*n*.123–124, 197*n*.161
Beauvoir, Simone de, 102

Beck, Clive, 158, 204n.76
behaviorism, emergence of, 87–89
belligerence, Dewey and, 66–67
Berlin, Isaiah, 76, 125
Berne, Eric, 203n.58
biological determinism, 30
biological potentiality (Gould), 2–3, 145
biological racism, 18, 22
biological Thatcherism, 89
blank-slate worldview, 1–2, 65, 74, 87, 91, 101
Boas, Franz, 85–86, 92–93

character
    education of, 145
    emergence of, 74
    sympathy and, 112
child-centered education, 54–55, 80, 81
children
    independence of, 41–42
    species barrier and, 160
closed instincts, 93–94
Colapinto, John, 192n.79
Coleridge, Samuel Taylor, 205n.84
communism, among indigenous societies, 30–31
communitarian philosophy, 59, 67–75
    acculturation process and, 35–36, 129, 147, 148–153
    concentric circles for curriculum, 158–161
    Dewey and, 59, 68–69
    liberal/communitarian debate and, 149–150
    MacIntyre and, 130–132, 149
    Midgley and, 109–111, 124–128, 132
compassion, as social instinct, 12
competition
    cooperation versus, 3–4, 5–7, 9, 19–24, 26–34, 49, 71–73
    Dewey and, 65–66
    Hobbes on, 2
    Kropotkin's attack on, 32–34
    natural selection and, 29–30
    "nature red in tooth and claw" (Tennyson), 19, 20, 22, 26–29, 54
    need for, 156
concentric circles of Darwinism, 19–22, 54, 158–161
conflict
    animal versus human conflicting desires, 12, 109–110
    in social instincts, 12, 14
    see also competition
conscience, 12
    development of, 14
    lack of, 13
consciousness, 21, 72, 88, 109–110, 114
constructivism, 150
cooperation
    competition versus, 3–4, 5–7, 9, 19–24, 26–34, 49, 71–73
    mutual aid and, 9, 30–33
courage, as social instinct, 9, 13
Cremin, Lawrence A., 34
cultural evolution
    natural evolution versus, 17–18, 34, 94–95
    in social Darwinism, 27
culture
    acculturation process and, 35–36, 129, 147, 148–153
    biology in manifestation of, 164
    in construction of human nature, 85–86, 137
    cultural memory and, 124
    cultural versus natural evolution, 34
    emergence from evolutionary nature, 94–95
    interface with nature, 31–32, 72–73
    interface with science, 3, 46–48, 56
    moral objectivity and, 114–118
    motivation and, 96–97
    as product of collective action, 110–111
    reality of evil and, 114–118
    science in, 154
    social instinct and, 67–75, 83–84
    socialization into, 35–36, 129, 147, 148–153
curiosity, wonder versus, 161–163
curriculum
    child-centered education and, 54–55, 80, 81
    concentric circles for, 158–161
    Darwinian, 153–161
    Dewey and, 79–81, 82
    Huxley and, 46–48
    Kropotkin and, 51–55
    Spencer and, 38–40, 41–42, 46
    usefulness/uselessness of, 155–156

Damasio, Antonio, 141, 200n.26, 201–202n.40

Darwin, Charles
on character, 112
on conflicting desires, 109–110
and cooperation versus competition, 3–4, 5–7, 9, 19–24, 29, 49
The Descent of Man, 8, 10–11, 15
Expression of the Emotions in Man and Animals, 11
gender differences and, 101–102
growth and, 61, 78
Huxley as bulldog for, 25–26
morality and, 8–16, 18, 44, 64–65, 94, 114–116, 120, 136, 144, 146–147, 169–170n.35–36, 194–195n.111
natural selection and. *see* natural selection
On the Origin of Species, 4–5, 7–8, 128, 168n.7, 168–169n.19
positive view of human nature, 3–24
science and, 37, 60
sexual selection and, 6–7, 16–17
social instincts and, 31, 54
struggle for survival and, 4–6, 168n.13, 168–169n.18–19
transition to Darwinism, 19–22
Victorian sensibilities and, 16–17, 22–23
worldview of, 123, 131, 154–155
Darwinism
altruism in, 9–10
as anti-religious worldview, 164
applications of, 23–24
Aristotelian worldview and, 4, 32, 130–132
behaviorism versus, 87–89
character of human beings in, 19–22
concentric circles of, 19–22, 54, 158–161
cooperation versus competition and, 3–4, 5–7, 9, 19–20, 23–24
curriculum under, 153–161
decline of, 85–86
development of, 1–24
of Dewey, 59, 60–62, 78–84, 130–134, 163–165
educational philosophy based on, 1–4, 12, 23–24, 78–79, 129, 130–140, 153–165
evolution versus intelligent design debates, 1–2
first circle of human nature, 19–22, 54, 158–161
first-generation, 2, 3–4, 5–6, 23–24, 25, 43, 56, 91, 101, 114, 122, 124, 127, 141
of Gould, 88, 90–91
implications of human nature in, 19–22
instinctual versus learned behavior, 14–16
of Midgley, 86, 89–90
natural world versus human beings in, 19–22
reemergence of, 86
second circle of human nature, 19–22, 54, 158–161
second-generation, 3–4, 56, 86, 89–90, 114, 122, 127–128
sexual selection and, 6–7, 16–17
survival of the fittest (Spencer) and, 5, 34–36, 39, 86, 122
third circle of human nature, 19–22, 54, 158–161
transition from Darwin, 19–22
waning of, 85–86
see also social Darwinism
Darwin Wars, 86
Dawkins, Richard
on poetry versus science, 123
as second-generation Darwinist, 114
"selfish" genes metaphor, 2–3, 7, 14, 86, 89–90, 122
social Darwinism and, 89–91
deconstructionist hermeneutics, 138
Degler, Carl, 85–86
dehumanization, 99, 117–118, 152
democracy, Dewey and, 75–78
Dent, Nicholas, 88
Descartes, René, 1, 162
Descent of Man, The (Darwin), 8, 10–11, 15
De Toqueville, Alexis, 125–126, 197n.162
DeWall, Franz, 189n.30
Dewey, John, 59–84
Aristotelian worldview and, 60, 61, 78, 130–132, 145–146, 152
change and, 60–62
communitarian ethic of, 59, 68–69
critics of, 81–83
curriculum and, 79–81, 82
Darwinism of, 59, 60–62, 78–84, 130–134, 163–165

democracy and, 75–78
educational philosophy of, 23, 78–84, 131–135, 137–140, 148–149
emotions and, 64–65
Experience and Education, 81, 188n.92
on feeling and action, 112–113
as first-generation Darwinist, 3–4, 124, 127
garden metaphor of, 28–29, 72–73, 134
growth and, 60–62, 63–65, 70–75, 78–83, 79
habits and, 15, 67–70, 81–82, 148–149
human nature and, 28–29, 57, 59–67, 70, 75–76, 77–78, 80–81, 133–134, 148–150
Huxley versus, 71–73
inequality and, 76–78
instincts and, 62–63, 66–68, 80–81, 93–94, 148–149
morality and, 64, 187n.84
negative liberty/positive liberty, 76
optimism of, 61, 75–76, 78
pedagogy and, 80, 83
pragmatic perspective of, 137–138
problem solving and, 53, 68–69, 82–83
progressive education and, 69–70, 80, 81–82, 180n.121
rationality and, 62–64, 68–69
Romantic tradition and, 79–80
social class differences and, 76–78
social instinct and, 67–75, 83–84
Spencer versus, 61
transcendental religious philosophy and, 62–64, 73–75
discrimination, defining, 102–103
Dubos, Rene, 174n.20

Eagleton, Thomas, 136
educational philosophy, 33–84, 129–165
blank-slate worldview and, 1–2, 65, 74, 87, 91, 101
child-centered education in, 54–55, 80, 81
cooperation versus competition in, 3–4, 12, 23–24
cultivating wonder in, 161–163
curriculum in. *see* curriculum
Darwinian perspective on, 1–4, 12, 23–24, 78–79, 129, 130–140, 153–165

of Dewey, 23, 78–84, 131–135, 137–140, 148–149
didactical implications of Darwinian approach, 161–163
education and selfish nature in, 2
emotion and reason in, 129, 140–147
habits in, 145–146, 148
of Huxley, 42–48, 49
of Kropotkin, 23–24, 48–57, 82–83, 132–133
metaphysical framework for, 163–164
of Midgley, 132–140
particularism/universalism in, 129, 148–153
pedagogy in, 40–42, 44–46, 53–54, 80
periods in, 140–144
rationality and, 129, 140–147
of Rousseau, 19, 79–81
socialization process and, 129, 148–153
of Spencer, 33–42, 178–179n.94, 181–182n.50
education of the virtues, 12–13
Egan, Kiergan, 34
Eibl-Eibesfeldt, Irenaus, 96
Eichmann in Jerusalem (Arendt), 117–118
emotions
absence of, 142
as both ends and means, 142–143
developing emotional intelligence, 156–158
Dewey and, 64–65
in educational philosophy, 129, 140–147
habits and, 152
human and animal, 11
in human nature, 21, 64–65
Kropotkin and, 46, 55, 64–65
moral reasoning and, 112–113, 145–147, 150–152
reason versus, 111, 129, 140–147
Spencer and, 38–39, 64–65, 176–177n.62
empathy, 12
Enlightenment thought, 125–126
environmental movement, 126
environmental studies, 52
escalator fallacy (Midgley), 91
Eskimos, communism among, 30–31
essentialism, 65–66, 135, 150
ethics
evolutionary versus human, 12, 16, 164–165

human nature and, 120–121
virtue and, 27–28
see also morality
ethology
animal behavior and, 95
differences between the sexes and, 101–104
human nature and, 92–93
eugenics movement, 18, 22, 86
evil, reality of, 3, 114–118, 146
Evolution and Ethics (Huxley), 43, 106
evolutionary theory
cultural versus natural evolution, 17–18, 27, 34, 94–95
ethics of, 43, 106, 164–165
great chain of being in, 7
human origins versus, 4–5
intelligent design versus, 1–2
motivation in, 92
mutual aid and. *see* mutual aid
natural selection in, 5–14, 85–86, 91–93
population growth and, 4–5
Evolution as a Religion (Midgley), 128, 191–192n.62
existentialism
De Beauvoir and, 102
Midgley and, 88–89, 126–127
Experience and Education (Dewey), 81, 188n.92
Expression of the Emotions in Man and Animals (Darwin), 11
extinction, fitness of culture and, 21–22

facts, values versus, 119–123, 130–131, 135–139, 154–155
fanatical individualism (Spencer), 27
fascism, 86
fatalism, 115
feminism, 100–106
De Beauvoir and, 102
"equality as sameness" strategy in, 102–103
pregnancy and, 102–105
rape legislation and, 103
fidelity, as social instinct, 9, 13
Firestone, Shulamith, 102
first-generation Darwinism, 2, 3–4, 5–6, 23–24, 25, 43, 56, 91, 101, 114, 122, 124, 127, 141, 174n.23

freedom
concept of, 87
liberal view of, 100–101
Midgley and, 197n.161
as negative ideal, 125
Freud, Sigmund, 2, 28
friendship, 142, 156

Gaia, 197n.165
Gallagher, Shaun, 150
Galton, Francis, 22
garden metaphor
of Dewey, 28–29, 72–73, 134
of Huxley, 28–29, 71–73, 134, 174n.20
of Midgley, 108
gender
differences between the sexes, 6–7, 16–17, 19, 100–106, 135
feminism and, 100–106
maternal instinct and, 12, 94, 102–105, 109–110
sexual selection and, 6–7, 16–17
in socialization process, 143
stereotypes concerning, 104–105, 135
genetics
natural selection and, 15, 85
sociobiology and, 90
genocide, 114
Ghiselin, M. T., 90
Gilligan, Carol, 193n.82
golden mean (Aristotle), 108–109
Gould, Stephen Jay, 168–169n.19, 175n.35
on biological potentiality, 2–3, 145
Darwinism of, 88, 90–91
Dawkins versus, 7
on Kropotkin, 25, 29
great chain of being, 7
great community (Dewey), 75
greatest happiness principle, 13–14
greenhouse gases, 117
group loyalty, 150–151
group selection theory, 9–10, 82
group welfare, as social instinct, 9–10
growth
Darwin and, 61, 78
Dewey on, 60–62, 63–65, 70–75, 78–83, 79
population, 4–5
see also garden metaphor
guilt, 12

habits
    dangers of, 69
    Dewey and, 15, 67–70, 81–82, 148–149
    in educational philosophy, 145–146, 148
    emotions and, 152
    evolutionary basis of, 132
happiness, 13–14
Heart and Mind (Midgley), 194n.94, 202n.45
heterosexuality, 135–136
Hickman, Larry A., 174n.20
hidden curriculum, 157
Hirst, Paul, 140, 141, 144, 149, 200n.25, 201n.36
Hobbes, Thomas, 1, 2, 91, 125, 162
home schooling movement, 41
homosexuality, 135–136, 138
Hughes, Judith, 103
humanistic education, 154
human nature
    animal behavior versus, 19–22, 136, 139
    anthropology and, 92–93
    blank-slate worldview versus, 1–2, 65, 74, 87, 91, 101
    concentric circles of, 19–22, 54, 158–161
    conflicting ideas about, 1–2
    consciousness in, 21, 72, 88, 109–110, 114
    cooperation versus competition in, 3–4, 5–7, 9, 19–24, 26–34, 49, 71–73
    cultural construction of, 85–86, 137
    in Darwinist thought, 1–24
    Dawkins on, 2–3
    Degler on, 85–86
    dehumanization versus, 99, 117–118, 152
    descriptive notions of, 19–20
    Dewey on, 28–29, 57, 59–67, 70, 75–76, 77–78, 80–81, 133–134, 148–150
    differences between the sexes and, 6–7, 16–17, 19, 100–106, 135
    emotions in, 21, 64–65
    essentialist view of, 65–66, 135, 150
    ethics and, 120–121
    evil and, 3, 114–118, 146
    evolutionary values in, 26, 90–91
    feminism and, 100–106
    first circle of, 19–22, 54, 158–161
    Gould on, 2–4
    Huxley on, 28–29, 65–66
    innate needs in, 3, 91–100
    is/ought dichotomy and, 118–123, 137–140, 143–144
    Kropotkin on, 3–4, 59, 146–150, 161, 175n.35
    learning in, 44
    Midgley on, 3–4
    morality in, 8–14, 54, 110–112
    natural world and, 133
    pessimistic views of, 28
    plasticity and, 2, 22, 86, 87–88
    of proximate versus ultimate ends, 39, 78, 91–92
    rationality in, 8–14, 21
    role of intelligence and, 97, 109, 111
    second circle of, 19–22, 54, 158–161
    sociability and, 109–111, 124–125
    socialization versus, 31
    third circle of, 19–22, 54, 158–161
Hume, David, 1, 11, 44, 106, 119–120, 142
Huxley, Thomas Henry
    artificial versus natural education and, 45–46
    curriculum and, 46–48
    as Darwin's bulldog, 25–26
    Dewey versus, 71–73
    dualist model of, 33–34
    educational philosophy of, 42–48, 49
    Evolution and Ethics, 43, 106
    as first-generation Darwinist, 43, 114
    garden metaphor and, 28–29, 71–73, 134, 174n.20
    natural philosophy of, 25–30, 33
    pedagogy and, 44–46
    progressive education and, 44–45, 56
    Romanes Lectures, 25–26, 43, 71
    Romantic tradition and, 43, 45
    science and, 46–48, 56–57

independence, of children, 41–42
individualism
    Midgley and, 105, 121
    sex-linked, 105
    Spencer and, 27, 35–36, 49
individual selection theory, 10
In Search of Human Nature (Degler), 85–86
instincts
    in animal behavior, 93–94, 109–110, 139
    conflict in, 12, 14
    Dewey and, 62–63, 66–68, 80–81, 93–94, 148–149
    habits and, 69–70
    Huxley and, 44

innate needs and, 3, 91–100
learned behaviors versus, 14–16
maternal, 12, 94, 102–105, 109–110
open versus closed, 93–94, 139
role of, 36
Spencer and, 38–39
teleological implications of having needs, 95–100
see also social instincts
intelligence
development of, 32
emotional, 156–158
evolution versus intelligent design debates, 1–2
mutual aid and, 31–32
role of, 97, 109, 111, 144–146
sociability in human, 109–111
see also rationality
is/ought dichotomy, 118–123, 137–140, 143–144

James, William, 63, 127
jealousy, need for, 156
Job, 163

Kant, Immanuel, 111, 170n.37
Keynes, John Maynard, 148
knowledge
meaning versus, 123
self-knowledge and, 113, 160
Koestler, Arthur, 96
Kosnik, Clare Maddott, 158, 204n.76
Kropotkin, Peter
attack on competition, 32–34
background of, 25–26, 29
curriculum and, 51–55
educational philosophy of, 23–24, 48–57, 82–83, 132–133
emotions and, 46, 55, 64–65
as first-generation Darwinist, 3–4, 5–6, 23–24, 25, 127, 174n.23
on human nature, 3–4, 59, 146–150, 161, 175n.35
inequality and, 49–51
morality and, 51–53, 54
Mutual Aid, 25–26, 30–33, 50–51, 174n.29, 175–176n.45
natural philosophy of, 25–26, 29–33
pedagogy and, 53–54
progressive education and, 54–55

rationality and, 63–65
Romantic tradition and, 30–31
science and, 51–53, 56–57

Lamarckian views, 15, 34, 55, 174n.23
language acquisition, 44, 67–69, 80–81
Lewis, C. S., 195n.123
liberalism
blank-slate worldview and, 2
goal of liberal education, 43, 47–48
liberal/communitarian debate, 149–150
libertarian movement, 126
Locke, John, 1, 2, 37
Lorenz, Konrad, 86, 96–97, 189n.12
love, 96, 142, 143, 153, 156, 161–162, 163

MacIntyre, Alastair, 130–132, 149, 199n.15
Maitland, F. W., 121
Malthus, Thomas, 4–5
Marx, Karl, 99
materialism, 62–63
maternal instinct
of bees, 94
moral instinct versus, 12, 109–110
pregnancy and, 102–105
Mayr, Ernst, 4
meaning
cultural memory and, 124
demise of, 130–131
knowledge versus, 123
Mendel, Gregor, 15, 85
metaphysics, educational philosophy and, 163–164
Midgley, Mary, 85–128, 138–160
Aristotelian worldview and, 4, 91–92, 97–98, 108–109, 116–117, 142, 145–146, 152, 163, 203n.62
Beast and Man, 86, 189n.12, 189n.24, 190n.35, 191n.45, 191n.60, 194n.95, 195n.123–124, 197n.161
communitarian philosophy of, 109–111, 124–128, 132
curiosity versus wonder and, 161–163
Darwinism of, 86, 89–90
differences between the sexes and, 17, 100–106, 135
educational philosophy of, 132–140
escalator fallacy and, 91
ethological theory and observation and, 92–93

Evolution as a Religion, 128, 191–192n.62
existentialism and, 88–89, 126–127
on facts versus values, 119–123
feminism and, 100–106
garden metaphor and, 108
Heart and Mind, 194n.94, 202n.45
individualism and, 105, 121
inequality and, 86–87
innate needs and, 3, 91–100
is/ought dichotomy and, 118–123, 137–140, 143–144
on moral objectivity, 114–118
open instincts and, 93–94
on pseudo-speciation, 150–151, 159–160
pseudo-speciation and, 150–151, 159–160
rationality and, 88–89, 203n.55
reason and emotion and, 129, 140–147
relativism and, 114–119, 121–122
as second-generation Darwinist, 3, 4, 56, 86, 122, 127–128, 170n.41
subjectivism of, 115, 121–122
whole person and, 96–97, 101–114, 124, 126
Wickedness, 191n.61, 193n.93, 194n.96, 195–196n.132
Mill, John Stuart, 100, 157, 204n.72
Moore, G. E., 119–121, 198n.5
morality
altruism and, 32
basis in social instinct, 9
biology and, 97–98
Darwin and, 8–16, 18, 44, 64–65, 94, 114–116, 120, 136, 144, 146–147, 169–170n.35–36, 194–195n.111
Dewey and, 64, 187n.84
emergence of, 8–16, 151–152
emotions and, 112–113, 145–147, 150–152
golden rule and, 11, 12–13, 108–109
in human nature, 8–14, 54, 110–112
Kropotkin and, 51–53, 54
lack of, 13
Midgley and, 114–118, 120–123
motivation and, 13, 14
mutual aid versus, 31
natural history of, 21
natural selection and, 8–14, 17
natural social motives and, 94
in the natural world, 12, 43–44
origins of mutual aid and, 29–30
pursuit of good life and, 142–143
rationality and, 8–14, 87–88, 141
reality of evil and, 114–118
selfishness versus, 13–15
Victorian sensibilities and, 16–17, 22–23, 35–37
see also ethics
motivation
in animal behavior, 87, 95–96
in building a whole life, 107
culture and, 96–97
emotions and, 142
evaluation of, 133–134
morality and, 13, 14
role in evolution, 92
whole person and, 96–97, 107–108
multicultural education, 159–160
multiculturalism, 84
murder, 114
Murdoch, Iris, 203n.62
music, 155
mutual aid
cooperation and, 9, 30–33
as evolutionary strategy, 130–132
Kropotkin and, 25–26, 30–33, 50–51
limitations of, 31–32
morality versus, 31
origins of, 29–30
Mutual Aid (Kropotkin), 25–26, 30–33, 50–51, 174n.29, 175–176n.45

Nagel, Thomas, 98
naturalistic fallacy (Moore), 119, 120
natural philosophy, 25–33
of Darwin, 3–4, 5–7, 9, 19–20, 29
of Huxley, 25–30, 33
of Kropotkin, 25–26, 29–33
of Spencer, 24, 25–27
natural selection
artificial selection versus, 18, 172n.64
competition and, 29–30
in evolutionary theory, 5–14, 85–86, 91–93
genetics and, 15, 85
as maximizing mechanism, 106–107
mechanics of, 85, 90–91
morality and, 8–14, 17
nature of, 5, 15
as proximate cause of behavior (Symons), 91–92
reproduction in, 6–7

role of, 6
  as a selfish process, 90–91
"nature red in tooth and claw" (Tennyson), 19, 20, 22, 26–29, 54
Nazi ideology, 22, 86, 117–118
needs
  aggression as, 96–97, 115, 156
  in animal behavior, 98
  innate needs in human nature, 3, 91–100
  repression of, 97
  teleological implications of having, 95–100
Neiman, Alvin, 130
neurobiology, rationalist model in, 141
Nineteenth Century, The (journal), 25–26
Noddings, Nel, 157–158
Nussbaum, Martha, 149, 159, 202n.49, 203n.55

obscurantism, Midgley and, 88
obsession, 116–117
On Aggression (Lorenz), 96–97
On the Origin of Species (Darwin), 4–5, 7–8, 128, 168n.7, 168–169n.19
open instincts, 93–94, 139
open school movement, 65, 81
optimism, of Dewey, 61, 75–76, 78
"other," 31–32

Paradis, James, 28
parental care
  evolution of, 96
  maternal instinct and, 12, 94, 109
  motive for nurturing and, 142
  Spencer and, 41, 42
particularism, educational philosophy and, 129, 148–153
pedagogy
  Dewey and, 80, 83
  Huxley and, 44–46
  interaction with real world and, 53–54
  Kropotkin and, 53–54
  Spencer and, 40–42
Pekarsky, Daniel, 62
Peters, Richard S., 141, 200n.26
physical education, 157
Piaget, Jean, 20
plasticity, 2, 22, 86, 87–88
Plato, 1, 93
play
  of animal and human babies, 93, 160–161
  differences between the sexes and, 104
pleasure, 12–13
poetry, science versus, 123, 153–156
population growth, evolutionary theory and, 4–5
positivist movement, 39–40, 56–57, 138
Postman, Neil, 130, 163
postmodern thought, 3
pregnancy, 102–105
problem solving
  Dewey and, 53, 68–69, 82–83
  Kropotkin and, 53
progressive education
  biases of, 35
  Dewey and, 69–70, 80, 81–82, 180n.121
  Huxley and, 44–45, 56
  Kropotkin and, 54–55
  Spencer and, 34, 42, 44–45, 56
proximate causes of behavior (Symons), 91–92
proximate ends, 39, 78, 91–92
pseudo-speciation (Midgley), 150–151, 159–160
punishment, Spencer and, 41

race theory, 86, 101, 135
rape, 103, 114
rationality
  critique of rationalist tradition, 141–142
  Dewey and, 62–64, 68–69
  educational philosophy and, 129, 140–147
  emotions versus, 111, 129, 140–147
  human ideals and, 2–3
  in human nature, 8–14, 21
  Midgley and, 88–89, 203n.55
  morality and, 8–14, 87–88, 141
  plasticity position and, 87–88
  prerational nature and, 140, 143
  social instinct combined with, 10–11
  Spencer and, 39–40
  of transcendental religious philosophy, 62–64
reciprocal altruism, 10
reductionism, 89–91
relativism, of Midgley, 114–119, 121–122
religion
  Darwinism as anti-religious worldview, 164
  science versus, 20

transcendental religious philosophy, 62–64, 73–75
repression, of needs, 97
reproduction, survival and, 6–7
responsibility, Midgley and, 115–117
Rockefeller, Steven, 64–65, 75
Roland Martin, Jane, 157–158
Romanes Lectures (Huxley), 25–26, 43, 71
Romantic tradition
   Dewey and, 79–80
   Huxley and, 43, 45
   Kropotkin and, 30–31
   Spencer and, 38
Rorty, Amelie Oksenberg, 167n.1
Rousseau, Jean-Jacques, 1, 19, 38, 79–81, 125
Ryan, Alan, 164–165, 182n.1

Scheffler, Israel, 140–141, 200n.26
science
   aesthetics versus, 39–40
   as central to religious education, 163
   curiosity versus wonder and, 161–163
   Darwin and, 37, 60
   explanatory power of, 3
   Huxley and, 46–48, 56–57
   interface with culture, 3, 46–48, 56
   is/ought dichotomy and, 137–140
   Kropotkin and, 51–53, 56–57
   in liberal education, 47–48
   Midgley and, 154–155
   natural laws versus intuition in, 36–38
   poetry versus, 123, 153–156
   in positivist movement, 39–40, 56–57, 138
   religion versus, 20
   Spencer and, 56–57
second-generation Darwinism, 3–4, 56, 86, 89–90, 114, 122, 127–128, 170n.41
secular humanism, 127
selfishness
   as basis of behavior in natural world, 14, 116
   morality versus, 13–15
   in natural selection, 90–91
   "selfish" genes metaphor (Dawkins), 2–3, 7, 14, 86, 89–90, 122
self-knowledge, 113, 160
Singer, Peter, 202n.50
slave labor, 117
Smith, Adam, 13, 170n.38
Sober, Eliot, 190n.25

sociability
   and human intelligence, 109–111
   and meaning of human life, 124–125
social change, ontological flexibility of human nature and, 86–87
social class
   Dewey and, 76–78
   Kropotkin and, 49–51
   Midgley and, 86–87
   social Darwinism and, 135
social Darwinism
   Dawkins and, 89–91
   Dewey and, 65–66
   differences between the sexes and, 101
   disappearance of Darwinism and, 85–86
   facts versus values and, 119, 121
   Midgley and, 89–91, 119
   nature of, 2, 14
   "nature red in tooth and claw" (Tennyson), 19, 20, 22, 26–29, 54
   reductionist, 89–91
   Spencer and, 27, 32, 33, 119
   see also Darwinism
social engineering, 18, 22, 87
social inequality
   Dewey on, 76–78
   Kropotkin and, 49–51
   Midgley on, 86–87
social instincts
   conflict in, 12, 14
   Darwin and, 31, 54
   Dewey and, 67–75, 83–84
   golden rule and, 11, 12–13, 108–109
   Kropotkin and, 49–50
   morality based in, 9–16, 18
   rationality combined with, 10–11
   sympathy as, 9, 12, 13, 15, 16, 150, 158, 159
socialization
   acculturation process and, 35–36, 129, 147, 148–153
   avoiding, 35–36
   as critical to curriculum, 156–158
   Dewey and, 68–69
   differences between the sexes and, 104–105
   educational philosophy based on, 129, 148–153
   gender differences in, 143
   human nature versus, 31

resisting, 99
Spencer and, 38
to wrong laws, 36–37
sociobiology
adaptationist agenda in, 90
altruism as central problem of, 90
genetic reproduction in, 90
Sociobiology (journal), 86
sociology
biology versus, 2–3
decline of Darwinism and, 85–86
differences between the sexes and, 104
speciation
Darwin and, 5
natural selection and, 5
pseudo-speciation (Midgley), 150–151, 159–160
Spencer, Herbert
curriculum and, 38–40, 41–42, 46
Dewey versus, 61
educational philosophy of, 33–42, 178–179n.94, 181–182n.50
emotions and, 38–39, 64–65, 176–177n.62
individualism and, 27, 35–36, 49
natural philosophy of, 24, 25–27
pedagogy and, 40–42
progressive education and, 34, 42, 44–45, 56
rationality and, 39–40
Romantic tradition and, 38
social Darwinism and, 27, 32, 33, 119
survival of the fittest and, 5, 34–36, 39, 86, 122, 168n.13
Victorian sensibilities and, 35–37
state
mandatory schooling and, 41
totalitarian government, 22, 86, 117–118
Stoics, 28, 46
struggle for existence, 71–75
struggle for survival, Darwin and, 4–6, 168n.13, 168–169n.18–19
subjectivism, of Midgley, 115, 121–122
survival of the fittest (Spencer), 5, 34–36, 39, 86, 122, 168n.13
sweatshops, 117
Symons, Donald, 91–92
sympathy
aggression versus, 95
as character trait, 112

morality based on, 17–18
as social instinct, 9, 12, 13, 15, 16, 150, 158, 159

technical education, 53
teleology
of Aristotelian worldview, 131–132
criticisms of, 64–65, 135–138
Dewey's rejection of, 64–65
feminism and, 100–106
implications of of having needs, 95–100
meaning of human life and, 123–128
of Midgley, 138–139
poetry and science in, 123, 153–156
reconstruction of, 130–131
ultimate versus proximate purposes and, 39, 78, 91–92
Tennyson, Alfred Lord, on "nature red in tooth and claw," 19, 20, 22, 26–29, 54
Tinbergen, Nikolaas, 86, 189n.12
totalitarian government, 22, 86, 117–118
transcendental religious philosophy, 62–64, 73–75
Tufts, James H., 183n.23

ultimate causes of behavior (Symons), 39, 78, 91–92
understanding, Midgley on, 123
universalism, educational philosophy and, 129, 148–153
utilitarianism, 13–14

values, facts versus, 119–123, 130–131, 135–139, 154–155
Victorian sensibilities, 16–17, 22–23, 35–37

Wallace, Alfred Russell, 168n.13
Warnock, Geoffrey, 118
Watson, John, 93
wholeness, Midgley and, 96–97, 101–114, 124, 126
wickedness, 114–118, 142
Wickedness (Midgley), 191n.61, 193n.93, 194n.96, 195–196n.132
Wilkins, Maurice, 205n.84
Wilson, David Sloan, 190n.25
Wilson, E. O., 86, 89–90, 91
wonder, curiosity versus, 161–163